Springer Theses

Recognizing Outstanding Ph.D. Research

Aims and Scope

The series "Springer Theses" brings together a selection of the very best Ph.D. theses from around the world and across the physical sciences. Nominated and endorsed by two recognized specialists, each published volume has been selected for its scientific excellence and the high impact of its contents for the pertinent field of research. For greater accessibility to non-specialists, the published versions include an extended introduction, as well as a foreword by the student's supervisor explaining the special relevance of the work for the field. As a whole, the series will provide a valuable resource both for newcomers to the research fields described, and for other scientists seeking detailed background information on special questions. Finally, it provides an accredited documentation of the valuable contributions made by today's younger generation of scientists.

Theses may be nominated for publication in this series by heads of department at internationally leading universities or institutes and should fulfill all of the following criteria

- They must be written in good English.
- The topic should fall within the confines of Chemistry, Physics, Earth Sciences, Engineering and related interdisciplinary fields such as Materials, Nanoscience, Chemical Engineering, Complex Systems and Biophysics.
- The work reported in the thesis must represent a significant scientific advance.
- If the thesis includes previously published material, permission to reproduce this must be gained from the respective copyright holder (a maximum 30% of the thesis should be a verbatim reproduction from the author's previous publications).
- They must have been examined and passed during the 12 months prior to nomination.
- Each thesis should include a foreword by the supervisor outlining the significance of its content.
- The theses should have a clearly defined structure including an introduction accessible to new PhD students and scientists not expert in the relevant field.

Indexed by zbMATH.

More information about this series at https://link.springer.com/bookseries/8790

Hun-ho Kim

Uniaxial Pressure Study of Charge Density Waves in a High-T_c Cuprate Superconductor

Doctoral thesis accepted by
the University of Stuttgart, Germany

Springer

Author
Dr. Hun-ho Kim
Solid State Spectroscopy
Max Planck Institute for Solid State Research
Stuttgart, Baden-Württemberg, Germany

Supervisor
Prof. Bernhard Keimer
Max Planck Institute for Solid State Research
Stuttgart, Germany

ISSN 2190-5053 ISSN 2190-5061 (electronic)
Springer Theses
ISBN 978-3-030-99900-1 ISBN 978-3-030-99898-1 (eBook)
https://doi.org/10.1007/978-3-030-99898-1

This Springer imprint is published by the registered company Springer Nature Switzerland AG
The registered company address is: Gewerbestrasse 11, 6330 Cham, Switzerland

Supervisor's Foreword

Quantum materials exhibit a large variety of phases with divergent macroscopic properties. A prime example is the phase diagram of layered cuprates, which includes charge-ordered and high-temperature superconducting phases in immediate proximity. In the quest for a microscopic description of the origin of these phases and their competition, the chemical disorder that is required to modify the doping level has proven to be a great complication. Hun-ho Kim performed an elegant series of resonant and non-resonant X-ray scattering experiments on the high-temperature superconductor $YBa_2Cu_3O_{6.67}$ under highly homogeneous uniaxial pressure and demonstrated the power of this method to modulate the interplay between superconducting and charge-density-wave (CDW) phases without introducing disorder and inhomogeneity. His work on uniaxial pressure tuning opens up many new perspectives for research on high-temperature superconductors and other quantum materials.

Hun-ho Kim first describes high-resolution non-resonant inelastic X-ray scattering experiments where uniaxial pressure was applied along one of the two inequivalent planar axes of the orthorhombic crystal structure. He highlights three key observations. First, the pressure enhances the intensity of a diffuse elastic X-ray signal originating from short-range-ordered CDWs in the CuO_2 planes that are weakly correlated perpendicular to the planes. Experiments performed in parallel by collaborators at the Max Planck Institute for Chemical Physics of Solids showed that the superconducting T_c is reduced under these conditions, consistent with a competition between both phases that was already noted in prior work in the absence of pressure. Second, he discovered a three-dimensional long-range ordered CDW under pressure that strains the CuO_2 layers by about 1%. The intensity of Bragg reflections characteristic of CDW order sets on sharply above T_c and disappears equally sharply upon cooling below T_c, signaling a strong thermodynamic competition between both phases. Third, the inelastic scattering signal showed striking evidence of optical phonon softening at the CDW transition, analogous to prior work on classical structural phase transitions. In my view, the observation of a three-dimensional long-range-ordered density wave state competing with superconductivity and of a corresponding soft mode can be regarded as a breakthrough in a research field that is well over three decades old.

Hun-ho Kim further reports results of a resonant scattering study under uniaxial pressure. Compared to the non-resonant experiments described above, these experiments were performed in a more versatile geometry that allowed the application of pressure to the two structurally inequivalent planar axes (*a* and *b*) of the orthorhombic structure. Again, there are three important new observations. First, the diffuse signal from short-range CDW order is enhanced by pressure along both *a* and *b*, but only if the pressure is applied perpendicular to the charge modulation direction. This observation implies uniaxial (rather than "checkerboard") charge ordering, thus conclusively resolving a puzzle that has been discussed intensively for several years. Second, three-dimensional long-range CDW order is only observed along the *b*-direction with pressure applied along *a*, and not vice versa, implying an important (indirect) role of the CuO charge reservoir layer in the crystal structure of $YBa_2Cu_3O_{6.67}$. Third, the photon energy dependence of the X-ray features indicative of two- and three-dimensional indicates charge modulations only within the CuO_2 layers, without direct participation of the CuO charge reservoir. These results will provide important guidance for further research on the microscopic origin of CDW order in the cuprates, its competition with superconductivity, and its influence on the macroscopic transport properties.

Stuttgart, Germany
March 2021

Prof. Bernhard Keimer

Abstract

The discovery of high-temperature superconductivity in strongly correlated copper oxides has triggered enormous theoretical and experimental efforts to unveil the microscopic mechanism behind superconductivity with a T_c that far exceeds that of the conventional superconductors known at the time. Elucidating the nature of the normal state of the cuprates, from which superconductivity arises, is the fundamental step to develop a theoretical framework that leads to a complete understanding of high-temperature superconductivity and possibly to the design of room-temperature superconductors. A vast amount of research shaped the currently known complex phase diagram of the cuprates, which is characterized by various electronic phases, each of which has been the subject of its own dedicated experimental endeavor. Despite almost 40 years of intense investigations, a consensus on their origin, nature and mutual relations has not been reached yet.

In the normal state of the cuprates, there have been many reports on lattice symmetry breaking of the electronic phase. Among these, density wave order has gained a considerable interest not only because of the early theoretical predictions of its existence [1–3] but also due to its intimate relation with other prominent phenomena, such as the pseudogap [4] and the anomalous suppression of superconductivity at 1/8 doping [5, 6]. In spite of extensive experimental studies dedicated to the charge density waves (CDWs) in cuprates, a static charge stripe phase was initially only observed in the single-layer 214 cuprates. The validity of the idea of universal stripe physics in CuO_2 planes [7] thus remains an open question, and alternative scenarios [8] were discussed with regard to the role of the low-temperature tetragonal phase of 214-cuprates in hosting stripes or a possible relation with the number of CuO_2 layers in a unit cell.

The recent technological development of resonant X-ray scattering allowed a major breakthrough in the study of CDWs in cuprates by proving their existence in all cuprate systems, thus revealing omnipresent charge ordering tendencies in cuprates. The overlap between the CDW stability range and the so-called underdoped regime, as well as the intertwined nature of the CDW that results in its competition with superconductivity, reconfirmed the importance of this electronic phase in describing the cuprate physics.

Here, I present inelastic X-ray scattering studies under uniaxial pressure performed on the underdoped cuprate $YBa_2Cu_3O_{6.67}$ ($p = 0.12$, $T_c = 65$ K) with incoming photon energy in the resonant (E = 931.3 eV, Cu-L_3 edge, RIXS) and non-resonant conditions (E = 17.794 keV, IXS). Uniaxial pressure is a unique tool to induce an anisotropy in the crystal structure in a clean way that subsequently affects the electronic structure. It has proven to be an effective tuning knob to manipulate the electronic ground state of correlated electrons systems, as shown in Sr_2RuO_4 [9, 10]. Its combination with X-ray scattering experiments that are the tool of choice to investigate the periodic charge modulations of CDWs in cuprates allowed us to explore a new phase space that was inaccessible in previous experiments.

The present doctoral thesis is composed of two parts. The first part of the thesis (Chaps. 1–3) provides background information on high-T_c cuprates and on the experimental methods employed in their investigation.

Chapter 1 introduces the history of the discovery and understanding of conventional superconductivity, which triggered later the search for higher superconducting critical temperatures, thereby giving birth to the new research field of unconventional superconductivity. The essential scientific ideas drawn from each stage of the research carried out on superconductivity are discussed.

Chapter 2 starts with explaining the basic properties of high-T_c cuprates, such as crystal and electronic structures. The underlying crystal structure determines the symmetry imposed on the electronic system, from which the various physical properties originate. The diverse electronic phases observed in the cuprate phase diagram are discussed in detail.

Chapter 3 provides explanations of the experimental methods that were utilized in the experiments. I discuss the reason why synchrotron radiation, rather than a conventional X-ray tube source, was used for our study as well as how X-rays interact with electrons. Crystal preparation is in general an important part of the pre-experiment procedure and particularly crucial for successful strain experiments. Details of how the samples were prepared and of the characterization of the strain effect are shown.

The second part of the thesis (Chaps. 4–6) shows results of experiments carried out using non-resonant and resonant inelastic X-ray scattering combined with uniaxial strain. The chapters are organized to explain the motivation, limitations, results and scientific insights of the experiments.

Chapter 4 presents the results from the IXS experiments on the underdoped cuprate $YBa_2Cu_3O_{6.67}$ ($p = 0.12$, $T_c = 65$K) with orthorhombic structure under uniaxial pressure. It shows the enhancement of the 2D-CDW along the b-axis under uniaxial pressure along the a-axis. In addition, it presents the first observation of the strain-induced, long-range 3D-CDW along the b-axis, while the phonon spectra display the complete softening of an optical phonon, thereby suggesting a soft phonon-driven mechanism for the formation of the 3D-CDW.

Chapter 5 shows the results of RIXS experiments on the same $YBa_2Cu_3O_{6.67}$ under uniaxial pressure. Our RIXS measurement reproduced the IXS data on the 3D-CDW along the b-axis, but did not find any evidence of a 3D-CDW along the a-axis. The main finding from RIXS is that the 2D-CDW is enhanced for uniaxial pressure applied perpendicular to the CDW, whereas the effect is considerably smaller under compression parallel to the CDW. This strongly anisotropic response cannot be accounted for within a two-dimensional checkerboard scenario. Instead, it can only be explained with the presence of domains having unidirectional character and modulation along the *a*- or *b*-axis, respectively. In addition, our measurement clearly shows the inequivalence of two orthogonal unidirectional 2D-CDWs even if the tetragonality of the CuO_2 plane is restored by *b*-axis compression. This and the absence of 3D-CDW order along the *a*-axis clearly suggest a crucial role of the oxygen chain layer in the CDW phenomenology.

Chapter 6 concludes the thesis by summarizing the key scientific achievements from the present studies. The experimental findings from the IXS and RIXS studies under uniaxial pressure and their implications are discussed with focus on future perspectives for studies of the CDW in all cuprate families.

Publications Related to This Thesis

1. **H.-H. Kim***, S.M. Souliou,*, M.E. Barber, E. Lefrançois, M. Minola, M. Tortora, R. Heid, N. Nandi, R.A. Borzi, G. Garbarino, A. Bosak, J. Porras, T. Loew, M. König, P.J.W. Moll, A.P. Mackenzie, B. Keimer, C. W. Hicks, and M. Le Tacon
"Uniaxial pressure control of competing orders in a high-temperature superconductor"
Science **362**, 6418 (2018).
2. **H.-H. Kim**, E. Lefrançois, K. Kummer, R. Fumagalli, N. Brookes, D. Betto, S. Nakata, M. Tortora, J. Porras, T. Loew, M. Barber, L. Braicovich, G. Ghiringhelli, A.P. Mackenzie, C. W. Hicks, B. Keimer, M. Minola, and M. Le Tacon
"Charge Density Waves in $YBa_2Cu_3O_{6.67}$ Probed by Resonant X-ray Scattering under Uniaxial Compression"
Physical Review Letters **126**, 037002 (2021).
* : authors with equal contribution

Acknowledgements

My doctoral studies in Stuttgart for the past four and a half years have been an invaluable experience in my life that shaped my identity as a scientist. It was an unique opportunity to work with excellent theoretical and experimental scientists.

First and foremost, I feel a profound gratitude to Prof. Bernhard Keimer for giving me this great opportunity to work in the exciting scientific project under the research environment that I am sure is not present anywhere in the world. His consistent guidance that spans the basic scientific discussion, valuable advice on setting research goals, as well as financial supports that enabled me to participate many synchrotron experiments and international research schools have been the main driving force for me to achieve my final goal.

I am very thankful to Prof. Matthieu Le Tacon who guided me to the intriguing research of uniaxial strain. Working with him, from my master studies to doctoral studies, has always been a great intellectual motivation and made me aim for higher goals in research. Especially, I enjoyed the scientific discussions that we had, where he always tried his best to give detailed answers from different points of view to my scientific questions.

I would like to sincerely thank my day-to-day supervisor Dr. Matteo Minola for the endless support and guidance that he provided during my entire doctoral studies. I learned so many things from him that are essential in research, including the very basics of synchrotron experiments and the true art of RIXS. Numerous experiments that we had together day and night, the discussions that we had, so many meals that we had together during the beamtime will never be forgettable. Throughout my doctoral studies, including the difficult times, his trust and mental support were indispensable for me to successfully finish my studies. It was a great luck to have him as my day-to-day supervisor.

I am also deeply grateful to Prof. Jörg Wrachtrup and Prof. Maria Daghofer for joining my thesis committee and evaluate my dissertation. Taking the exam from them was a challenging, but a unique, intellectually exciting experience.

My Ph.D. project was largely about developing a new experimental environment and therefore required substantial technical supports from so many people. I would like to thank Benjamin Bruha, Michael Schulz, Heiko Uhlig and Peter Wochner for helping me to modify experimental setups for strain experiments in MPI. I would also thank Gaston Garbarino for his technical supports in establishing the strain environment at the ID27 and ID28 of ESRF. I am very grateful to Manfred Ohl, Peter van der Linden for designing and modifying the RIXS chamber for the strain experiments at the ID32 of ESRF. I would also thank Mark Barber, Clifford Hicks and Prof. Andrew Mackenzie for all the extensive scientific and technical aids that were absolutely necessary for successful strain experiments. I appreciate the service groups in MPI-FKF, including the low-temperature service, crystal preparation service, crystal growth group, mechanical work shop, electric shop for their services. I particularly thank Barbara Baum for preparing nice crystals, and Kathrin Guelck for fabricating sample plates quickly.

Physically, mentally and technically challenging experiments in synchrotrons could not have been successfully done without the help of the outstanding colleagues. I owe Michaela Souliou a lot of thanks for the great success of the experiments at the ID28. I would also like to thank Emilie Lefrançois for her essential on-site contribution in the ID32 experiments. I feel deeply grateful to Roberto Fumagalli, Michele Tortora, Davide Betto, Suguru Nakata, Lucio Braicovich for their great contributions and the exciting experiences that we had in various strain beamtimes. I am very thankful to Kurt Kummer and Nick Brookes for always providing a great experimental environment for RIXS.

I would like to thank the formal and present members of our department for all the scientific discussions and help, as well as the lunch and coffee time that we had: Many thanks to Eva Benckiser, Joel Bertinshaw, Martin Bluschke, Alexander Boris, Sungkyun Choi, Robert Dawson, Christopher Dietl, Katrin Fürsich, Hlynur Gretarsson, Laura Guasco, Namrata Gurung, Matthias Hepting, Marc Höppner, Sewon Jung, Giniyat Khaliullin, Prof. Bumjoon Kim, Timofei Larkin, Humei Liu, Yi Lu, Simon Mayer, Roberto Ortiz, Daniel Pröpper, Rebecca Pons, Ksenia Rabinovic, Padma Radhakrishnan, Jorge Sauceda, Cissy Suen, Nakheon Sung, Hakuto Suzuki, Heiko Trepka, Kentaro Ueda, Lichen Wang, Meng Wu, Zichen Yang, Peter Yordanov, Junbang Zeng, Valentin Zimmermann. I would like to give special thanks to Juan Porras and Toshinao Loew for giving their fruitful know-hows in crystal preparation. I would also like to thank my former and present office mates: Daniel Putzky, Maximilian Krautloher, Xiatong Shi, Friederike Wrobel. Especially I would like to appreciate Daniel Putzky for helping me writing a German version of the abstract.

Interactions with Koreans in the institute were important parts of my life in Stuttgart. I had great memories of all the dinners with Sungkyun Choi, Gideok Kim, Minu Kim. I remember many exciting and fun moments with Gee Yeong Kim, Insook Kim, Junhong Na, Min-Jae Kim, Hyun Gyu Kim, Dong Yeong Kim, Hyunah Kwon, EunJin Choi, Mina Jung. I would also thank Prof. Ambrose Seo for having many interesting conversations and providing valuable advice in research and life. Lunch time in the canteen was full of funny conversations with many Koreans. I would like to give many thanks to Hyeon-Ho Jeong, Prof. Chang Uk Jung, Soon Jung Jung, Hoon Kim, Hyojeong Kim, Hojin Lee, Hyosang Lee, Yunwoo Lee, Saekwang Nam,

Bong Gyu Shin, Jaeho Shin, Donghoon Son, Kwanghyo Son, Sukho Song. Outside MPI, I owe great sources of fun to Daesik Kwak, Seryun Jang, Jiyoung Kim.

I would like to thank Kyungmi Lim for providing all the assistance to me. Especially during the times of Corona virus at the end of my study, her endless support was essential in overcoming the difficulty that I encountered while I was writing my dissertation. Without her consistent encouragement, I would not have successfully finished my study.

I would like to end this long acknowledgement section by giving my biggest appreciation to my parents, who have supported me in every aspect. Their presence has been the greatest mental base for me to go through all the challenges that I had during my doctoral studies.

Contents

1 Introduction 1
References 4

2 High Temperature Cuprate Superconductors 5
2.1 Crystal Structure 5
2.1.1 $YBa_2Cu_3O_{6+x}$ 6
2.2 Electronic Structure 9
2.3 Phase Diagram and Phenomenology 13
2.3.1 Antiferromagnetism 14
2.3.2 Superconductivity 17
2.3.3 Pseudogap 21
2.3.4 Charge Density Waves 26
References 35

3 Experimental Methods 41
3.1 Synchrotron Radiation 41
3.2 X-Ray Scattering Technique 45
3.2.1 Thomson Scattering 46
3.2.2 Dynamical Structure Factor 47
3.2.3 Experimental Setup 49
3.3 Uniaxial Strain Technique 53
3.3.1 Piezoelectric-Based Strain Device 53
3.3.2 Characterization 55
3.4 Crystal Preparation 59
3.4.1 Annealing 59
3.4.2 Detwinning 61
3.4.3 Preparation for Strain Experiment 62
References 62

4 CDW in $YBa_2Cu_3O_{6.67}$ Under Uniaxial Pressure: Non-resonant Inelastic X-Ray Scattering 65
4.1 Background 65

4.1.1 Electron-Phonon Interaction and CDW in Cuprates 65
4.1.2 Lattice Pressure Studies of YBCO 65
4.1.3 Uniaxial Pressure as a New Perturbation Method 66
4.2 Experiment 67
4.2.1 Typical IXS Spectrum 69
4.2.2 Energy Resolution 70
4.2.3 Momentum Resolution 70
4.3 Results 71
4.3.1 Enhancement of 2D-CDW Under Uniaxial Pressure 71
4.3.2 Emergence of 3D-CDW Under Uniaxial Pressure 73
4.3.3 Complete Softening of an Optical Phonon at the Onset of the 3D-CDW 76
4.4 Discussion 78
4.4.1 Comparison with Magnetic-Field-Induced 3D-CDW 78
4.4.2 Comparison Between 2D- and Uniaxial-Pressure-Induced 3D-CDW 79
4.5 Conclusions 79
References 80

5 CDW in $YBa_2Cu_3O_{6.67}$ Under Uniaxial Pressure: Resonant Inelastic X-ray Scattering 83
5.1 Background 83
5.2 Experiment 84
5.2.1 Typical RIXS Spectrum 85
5.2.2 Energy Resolution 87
5.2.3 Momentum Resolution 87
5.3 Results 88
5.3.1 Confirming the Uniaxial Nature of CDW in Cuprates 90
5.3.2 Absence of the 3D-CDW Under *b*-axis Compression 91
5.3.3 Same Resonance Profile of 2D- and 3D-CDW 92
5.3.4 Anisotropic Response of the 2D-CDW Under Uniaxial Pressure 93
5.4 Discussion 96
5.4.1 Direct Versus Indirect Effects of Uniaxial Pressure 96
5.4.2 *b*-axis Compression Does Not Restore the C_4-Symmetry: Role of the CuO Chain Layer 97
5.4.3 Anisotropic and Symmetric Response of the 2D-CDW to Uniaxial Pressure 98
5.4.4 Uniaxial Pressure Enhances the CDW Perpendicular to the Pressure Direction 99
5.4.5 Why the 3D-CDW is Observed Only Along the *b*-axis 99
5.5 Conclusions 101
References 102

6 Conclusion 103

Definitions and Acronyms

Definitions

Soft X-rays	X-rays with the energy $E < 1$ keV
Tender X-rays	X-rays with the energy $E = 1$–5 keV
Hard X-rays	X-rays with the energy $E > 5$ keV

Acronyms

2D-BZ	Two-dimensional Brillouin zone
a-CDW	Charge density waves with the modulation along the a-axis
ARPES	Angle-resolved photoemission spectroscopy
b-CDW	Charge density waves with the modulation along the b-axis
BW	Bandwidth
CDW	Charge density waves
DFT	Density functional theory
ESRF	European synchrotron radiation facility
FWHM	Full width at half maximum
HWHM	Half width at half maximum
ID	Insertion device
IXS	Inelastic X-ray scattering
LDA	Local density approximation
NMR	Nuclear magnetic resonance
PGM	Plane grating monochromator
QPI	Quasiparticle interference
RIXS	Resonant inelastic X-ray scattering
RXS	Resonant X-ray scattering
STM	Scanning tunneling microscopy
STS	Scanning tunneling spectroscopy

VLS	Variable line spacing
YBCO	$YBa_2Cu_3O_{6+x}$

Chapter 1
Introduction

The history of conventional superconductor research, from its discovery to the complete understanding of its microscopic mechanism, is not only an interesting story in itself, but also provides a good research model to tackle the problem of unconventional superconductivity in cuprates. Superconductivity was first observed in resistivity measurements of the metal mercury below T = 4.2 K by Heike Kamerlingh Onnes in 1911. This intriguing many-body phenomenon opened a new field of research and took almost 50 years to be properly understood when Bardeen, Cooper and Schriefer (BCS) proposed the superconducting pairing mechanism via the electron-phonon interaction in 1957.

It is not surprising that it took so many years until the microscopic, quantum-mechanical description of the BCS theory appeared. An understanding of the quantum nature of the electrons did not exist at the time of the discovery of superconductivity. The classical description of the electron transport properties in metals, the so-called Drude model, was proposed in 1900 by Paul Drude, and subsequently refined by Hendrik Lorentz over the next decade. The quantum-mechanical formulation of the theory of non-interacting electrons in solids was first proposed by Felix Bloch in 1928. The isotope effect, represented as the relation $M^{1/2}T_c = const$, was observed in 1951, and the electron-phonon interaction was immediately proposed as a pairing glue. Theoretical attempts to adequately describe the electron-phonon interaction began by Fröhlich [1], followed by Bardeen and Pines [2]. It was only in 1957 that the microscopic theory of superconductivity was finally published [3, 4] and explained all the key features of superconductivity: the energy gap $2\Delta = 3.5k_B T_c$, the isotope effect $M^{1/2}T_c = \text{const}$, the infinite conductivity, the Meissner effect, and the thermodynamic properties.

The triumph of the BCS theory relied on a lot of theoretical and experimental research efforts, each of which clarified and solved the individual puzzles of the whole physical picture. Specifically, a correct description of the normal state of

H. Kim, *Uniaxial Pressure Study of Charge Density Waves in a High-T_c Cuprate Superconductor*, Springer Theses,
https://doi.org/10.1007/978-3-030-99898-1_1

the superconductor from which superconductivity arises, several experiments that pinpoint the key features of superconductivity, and a proper mathematical tool to describe the pairing interaction were the key to the success of the BCS theory of superconductivity.

Boosted by the achievement of the BCS theory, immense research efforts were devoted to enhance the superconducting transition temperature. The equation for the superconducting transition temperature in the BCS theory

$$T_c = 1.14\Theta_D \exp\left(-\frac{1}{N(0)V}\right)$$

implies that high T_c is favoured by a high Debye temperature Θ_D, a large density of states at the Fermi level $N(0)$ and a strong electron-phonon coupling constant V. However, separate manipulation of these parameters is not easy in real materials, since they are coupled with each other. Increasing the carrier concentration $N(0)$ can result in a strong screening, which in turn leads to a weak electron-phonon interaction V. Aiming for a stronger electron-phonon coupling can, however, induce structural instabilities.

Metallic hydrogen is an extreme case of increasing the average phonon energy Θ_D, as suggested by Ashcroft [5]. The theory of metallic hydrogen under high pressure predicts a Debye temperature $\Theta_D = 3500\,\text{K}$ and $N_0V \sim 0.25$. This idea was recently realized in H_3S($T_c = 203\,\text{K}$) [6] and LaH_{10}($T_c = 260\,\text{K}$) [7] under high pressure of 150–200 GPa, comparable to the pressure of the Earth's core.

Even before the discovery of high-T_c cuprates, superconductivity in oxides was not unprecedented. Nb-doped $SrTiO_3$ ($T_c = 1.2\,\text{K}$), $Ba_{1-x}Pb_xBiO_3$ ($T_c = 13\,\text{K}$), and $Li_{1+x}Ti_{2-x}O_4$ ($T_c = 14\,\text{K}$) were found to be superconducting, albeit being far from high-temperature superconductivity. When Bednorz and Müller, who discovered high-T_c in 214 cuprates in 1986, began to search for new oxide superconductors, they aimed for increasing the electron-phonon interaction V by means of the so-called polaron formation.

They were inspired by the model of Jahn–Teller polarons for intermetallic narrow-band compounds proposed by Höck et al. [8], who claimed that Jahn–Teller polarons are formed when the Jahn–Teller distortion energy E_{JT} and the bandwidth t are comparable. In this situation, one of the degenerate bands susceptible to the Jahn–Teller distortion is strongly coupled with the structural deformation of the corresponding symmetry and, as a result, the composite of the electron and the local structural distortion moves freely through the lattice.

Examples of Jahn–Teller ions are Fe^{4+}, Ni^{3+} and Cu^{2+} that have a single electron occupying one of the degenerate d-bands in a cubic crystalline environment. Their first attempt in the La-Ni-O system with tuning the Jahn–Teller distortion energy by Al-doping did not succeed, and only induced a metal-semiconductor transition. It was the Ba-La-Cu-O system with the perovskite structure where they found superconductivity with $T_c \sim 30\,\text{K}$. This motivated extensive searches for copper oxide superconductors that subsequently led to the discovery of $YBa_2Cu_3O_{6+x}$, the first superconductor with T_c exceeding the liquid nitrogen temperature and other families of cuprates, such as Bi-, Tl-, Hg-based compounds, which recorded the highest T_c at ambient pressure (135 K) and T_c under pressure (165 K).

It was soon realized, however, that the Jahn–Teller effect is largely quenched by the strongly distorted crystalline environment in the layered structure of the cuprates, and that Jahn–Teller physics is thus unlikely to hold the key to the origin of high-T_c superconductivity. Rather, doped holes with ligand oxygen character in the CuO_2 planes determine the physical properties of the high-T_c cuprates, resulting in a complex phase diagram that shows various electronic phases which are up to this date not fully understood. One of the main difficulties is the lack of an adequate theoretical description of the unconventional, non-Fermi-liquid normal state. This gives a different starting point in the study of superconductivity, compared to the case of conventional superconductivity, where the normal state is well described by Bloch's single-electron model.[1]

Among the various electronic phases in the normal state of the so-called underdoped regime (where the density of mobile carriers is smaller than the one required for optimal T_c), charge order has gained considerable interest in relation to the early mean-field studies of Hubbard models that predicted incommensurate uni-directional charge/spin density waves [10, 11] and the subsequent experimental observation of such states in $La_{1.6-x}Nd_{0.4}Sr_xCuO_4$ [12]. Despite the substantial amount of research, experimental evidence of the charge order had been available only in the 214-family of cuprates for almost a decade.

A renewed interest was found following the discovery of charge density waves (CDWs) in other cuprate families. After the first discovery of charge order in $RBa_2Cu_3O_{6+x}$(R = Nd, Y) and thanks to fast-paced technical improvements, especially of resonant x-ray scattering techniques, it was subsequently observed in all cuprate systems. It is now generally accepted that the charge ordering tendency is an intrinsic property of underdoped cuprates. Charge order has been extensively studied in various cuprate compounds using different perturbation methods, such as temperature, magnetic field, hydrostatic pressure, and chemical doping. Nevertheless, the origin of the charge order in cuprates still remains unclear.

The role of experimentalists and experiments is to provide new pieces of information beyond the theoretical predictions and impose boundaries on future theoretical formulations. Compared to conventional perturbation methods, the uniaxial strain technique used in the present thesis work is, despite its long history, a rapidly developing technique in the field of strongly correlated electron systems. The basic idea of the uniaxial strain is simple. One needs two parallel anvils to apply pressure on the sample between them. The technical difficulty arises when it is combined with extreme conditions such as a cryogenic environment that the low-energy-scale phenomena of correlated electron systems often require. Moreover, given the high sensitivity of a strongly correlated electron system to external perturbations, homogeneity of the applied pressure throughout the sample volume is of utmost importance.

[1] To be precise, it is described by Fermi liquid theory, but this theory was proposed only in 1957 [9].

Uniaxial pressure studies on the unconventional superconductor Sr_2RuO_4 have recently led to a breakthrough. A newly-developed piezoelectric-based strain device enabled the application of uniaxial pressure of 1–2 GPa at cryogenic temperatures $T < 2$ K in a highly homogeneous fashion. The researchers reliably tuned the superconducting T_c under uniaxial strain [13] and induced a Lifshitz transition in Sr_2RuO_4 at the uniaxial strain of $\sim\!-0.6\%$ [14]. Exploiting this technical development and a modified version of this device, we performed non-resonant and resonant inelastic x-ray scattering experiments on the underdoped cuprate $YBa_2Cu_3O_{6.67}$ ($p = 0.12$, $T_c = 65$ K) under uniaxial pressure. We studied how the anisotropy imposed on the crystal structure affects the CDW phenomenology in terms of its correlation length in the copper-oxide plane and out of plane. The relation between the CDW and phonons was also investigated, inspired by the strong momentum-specific electron-interaction reported in previous studies [15, 16].

References

1. Fröhlich H (1950) Phys Rev 79:845–856. https://doi.org/10.1103/PhysRev.79.845. Accessed 5 Sept 1950
2. Bardeen J, Pines D (1955) Phys Rev 99:1140–1150. https://doi.org/10.1103/PhysRev.99.1140. Accessed 4 Aug 1955
3. Bardeen J, Cooper LN, Schrieffer JR (1957) Phys Rev 106:162–164. https://doi.org/10.1103/PhysRev.106.162. Accessed 1 April 1957
4. Bardeen J, Cooper LN, Schrieffer JR (1957) Phys Rev 108:1175–1204. https://doi.org/10.1103/PhysRev.108.1175. Accessed 5 Dec 1957
5. Ashcroft NW (1968) Phys Rev Lett 21:1748–1749. https://doi.org/10.1103/PhysRevLett.21.1748. Accessed 26 Dec 1968
6. Drozdov AP et al (2015) Nature 525:73–76
7. Drozdov AP et al (2019) Nature 569:528–531
8. Höck K-H, Nickisch H, Thomas H (1983) Helvetica Phys Acta 237–243
9. Landau LD (1957) Soviet Phys JETP 3:6
10. Zaanen J, Gunnarsson O (1989) Phys Rev B 40, PRB:7391–7394
11. Poilblanc D, Rice TM (1989) Phys Rev B 39:9749–9752. https://doi.org/10.1103/PhysRevB.39.9749. Accessed 13 May 1989
12. Tranquada JM et al (1995) Nature 375
13. Hicks CW et al (2014) Science 344:283–285. https://doi.org/10.1126/science.1248292. https://science.sciencemag.org/content/344/6181/283.full.pdf
14. Steppke A et al (2017) Science 355:148
15. Le Tacon M et al (2014) Nat Phys 10:52–58
16. Blackburn E et al (2013) Phys Rev B 88:054506. https://doi.org/10.1103/PhysRevB.88.054506. Accessed 5 Aug 2013

Chapter 2
High Temperature Cuprate Superconductors

2.1 Crystal Structure

After the first discovery of high temperature superconductivity in the La-Ba-Cu-O system [1], a superconducting T_c that far exceeds the BCS limit was subsequently observed in various forms of copper oxides, as shown in Fig. 2.1a. Although differences in their chemical compositions and crystal structures exist between different cuprate compounds, they commonly have a layered structure consisting of CuO_2 planes (Fig. 2.1b) sandwiched by charge reservoir layers.

The CuO_2 planes, commonly found in all cuprate compounds, are the stage for most of the interesting cuprate physics, as will be discussed in the next section. The number of CuO_2 planes in a given unit cell of the known cuprate compounds can go from one (La-based 214-compounds) to three or more (Bi, Tl, Hg-based compounds).

A chemical change of the charge reservoir can affect the doping level of the CuO_2 planes, either by adding (electron-doping) or removing (hole-doping) electrons. The methods to chemically manipulate the charge reservoir include heterovalent substitution of the rare-earth ions (Sr-substitution in La_2CuO_4), introduction of interstitial oxygens ($La_2CuO_{4+\delta}$, $Bi_2Sr_2CaCu_2O_{8+\delta}$), and adding oxygens at crystallographically allowed lattice sites (for $YBa_2Cu_3O_{6+x}$). Although the phase diagram of the electron-doped cuprates looks similar to that of the hole-doped counterpart, the description of the electron doping case requires a different theoretical formulation, since the character of the doped carriers is different (Cu-3d character in electron-doped cuprates versus O-2p character in hole-doped cuprates). We will focus on the hole-doped case throughout the thesis, which has been extensively studied for the last 35 years.

Since the experiments presented in the thesis were carried out specifically for the $YBa_2Cu_3O_{6+x}$ system, a detailed explanation of the crystal structure of $YBa_2Cu_3O_{6+x}$ will be provided in the next section.

H. Kim, *Uniaxial Pressure Study of Charge Density Waves in a High-T_c Cuprate Superconductor*, Springer Theses,
https://doi.org/10.1007/978-3-030-99898-1_2

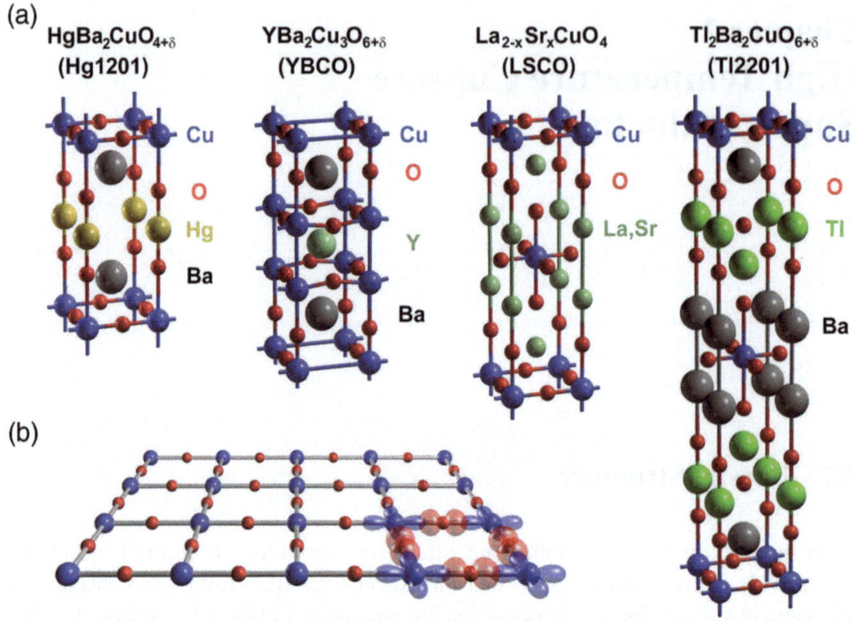

Fig. 2.1 **a** Crystal structures of different cuprate compounds. Detailed crystal structures vary depending on the hole doping levels. For example, interstitial oxygens induce holes on the CuO_2 planes in the Hg1201 compounds. **b** Illustration of the CuO_2 plane with the Cu-3$d_{x^2-y^2}$ and O-2p orbitals. Reprinted figure from [2]. Copyright (2013) National Academy of Sciences

2.1.1 $YBa_2Cu_3O_{6+x}$

The unit cell of the $YBa_2Cu_3O_{6+x}$ (YBCO) crystal is illustrated in Fig. 2.2. It is a layered structure consisting of two CuO_2 planes intercalated by CuO chain layers, which act as charge reservoir, and by BaO insulating layers and an Y layer. There are two different positions for Cu atoms in the unit cell: Cu(1) and Cu(2). The Cu(1) atoms are linked to the O(1) atoms that form the CuO chain layer. The O(1) sites in the CuO chain layer are empty in the parent compound ($x = 0$) and completely occupied in the fully-oxygenated YBCO ($x = 1$). At intermediate oxygen contents ($0 < x < 1$), the O(1) sites are partially filled, resulting in various oxygen superstructures in the CuO chain layer.

The Cu(2) atom in the CuO_2 plane is surrounded by a pyramid of the O(2), O(3) and O(4) atoms. The distances between Cu(2) and neighboring oxygens are, however, not equal. The bond length of Cu(2)-O(4) (∼2.3 Å) is much larger than the bond length of Cu(2)-O(2)/O(3) (∼1.9 Å), thereby yielding the two-dimensional physics of the CuO_2 plane. Anisotropic Cu-O bond lengths are commonly observed in high-T_c cuprates. The structural refinement studies [3–5] show that the in-plane Cu-O-Cu bond is not straight (180°), but slightly bent (∼165°, in-plane buckling).

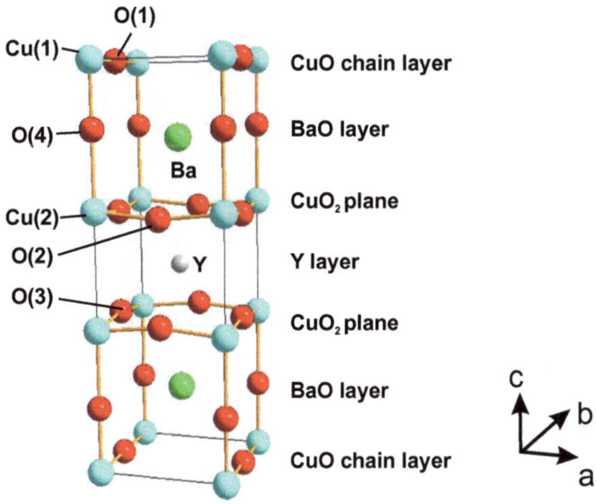

Fig. 2.2 Crystal structure of $YBa_2Cu_3O_{6+x}$. Here, the case of x = 1 is shown

The Y layer lies at the center of the unit cell that separates the two CuO_2 planes. It exhibits a localized, ionic character in the entire doping range, as demonstrated by its constant bond valence sum of 2.9 upon hole doping [4] and therefore, it merely plays the role of stabilizing the structure. The insulating BaO layer is present between the CuO chain layer and the CuO_2 plane.

YBCO undergoes a structural transition from the tetragonal ($a = b = c, \alpha = \beta = \gamma = 90°$) to the orthorhombic ($a \neq b \neq c$, $\alpha = \beta = \gamma = 90°$) crystal structure as the additional oxygen content x increases. Figure 2.3 shows the change of the lattice constants of $YBa_2Cu_3O_{6+x}$ as a function of $\delta = 1 - x$, measured by neutron powder diffraction [3]. The undoped compound ($x = 0$, $\delta = 1$) has a tetragonal structure

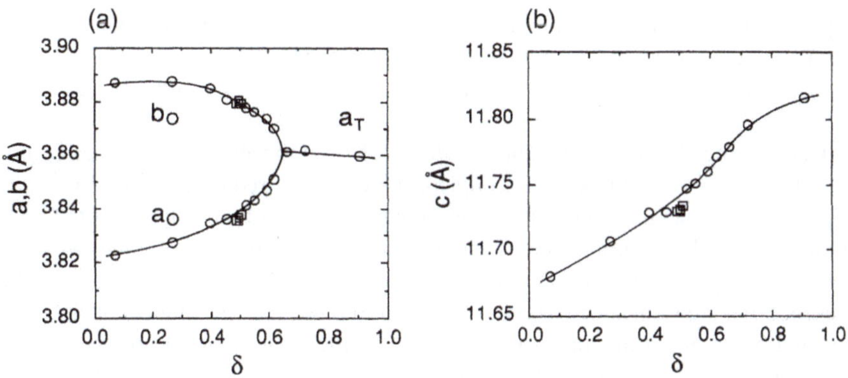

Fig. 2.3 Change of the **a** a, b-lattice constants **b** c-lattice constant as a function of δ in $YBa_2Cu_3O_{7-\delta}$. Reprinted figure with permission from [3]. Copyright (1990) by the American Physical Society

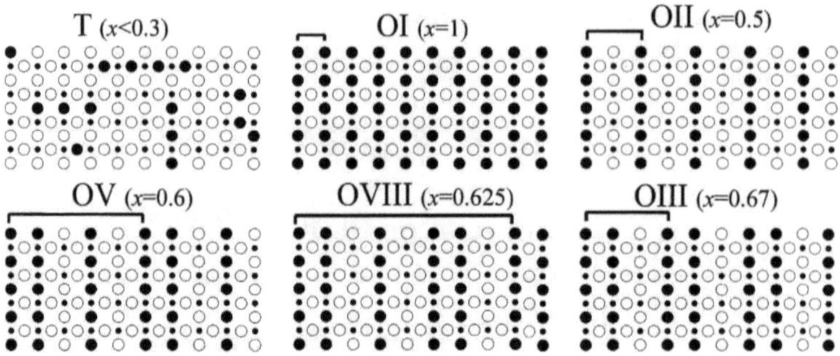

Fig. 2.4 Experimentally observed oxygen superstructures of YBCO. Filled(empty) circles represent occupied(unoccupied) O(1) sites. T is the disordered state of oxygens in the tetragonal phase. OI-OVIII refer to the Ortho-I to Ortho-VIII oxygen orders. Reprinted figure from [6]

with lattice constants of $a = b = 3.86\,\text{Å}$, $c = 11.84\,\text{Å}$ at room temperature. The tetragonal phase remains upon further increase of x, only slightly expanding the lattice by the addition of oxygens in the CuO chain layer, until it reaches $x \sim 0.35$. Above $x \sim 0.35$, two distinct lattice constants are observed and the orthorhombic distortion ($\frac{2(b-a)}{b+a}$) increases as the oxygen content $x = 1 - \delta$ increases.

In the orthorhombic phases ($x > 0.35$), long-range oxygen superstructures are observed, which are not present in the tetragonal phase ($x < 0.35$). These are characterized by distinct patterns of empty and filled O(1) sites along the a-axis and filled O(1) sites along the b-axis. Experimentally observed oxygen superstructures are shown in Fig. 2.4. Depending on the oxygen doping level x, various patterns of oxygen orders with a different periodicity exist, namely from Ortho-I (period of one lattice constant) to Ortho-VIII (period of eight lattice constants) orders.

The appearance of long-range oxygen order in the CuO chain layer plays a critical role in the charge transfer from the CuO_2 plane to the CuO chain layer. This was experimentally demonstrated by the observation of different superconducting T_c in the $YBa_2Cu_3O_{6.50}$ single crystals with ortho-I and ortho-II oxygen orders [7]. This can be explained in a qualitative level, as discussed below and in the review paper [8].

Cu(1) in the CuO chain layer has a valence of 1+ in the absence of oxygen chains ($x = 0$). When one O(1) site is occupied, the electrons required to make O^{2-} are supplied by the neighboring Cu(1) ions, converting each of the Cu(1) ions to Cu^{2+}. As a result, it does not induce any charge transfer from the CuO_2 plane, which is approximately the situation occuring at the lower oxygen contents $x < 0.35$. In the presence of n adjacent added oxygens in the CuO chain layer, the number of electrons donated by neighboring copper ions is $n + 1$, while it takes $2n$ for all the oxygens to become O^{2-}. For large n, the 'shortage' of the charges in the CuO chain layer is resolved by charge transfer from the CuO_2 plane, introducing holes in the CuO_2 plane (Fig. 2.5).

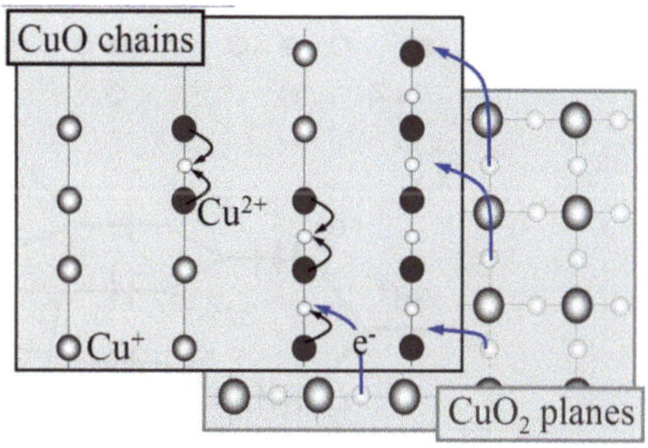

Fig. 2.5 Illustration of how full chains of oxygens results in the charge transfer from the CuO_2 plane. Short segments of the oxygen chains are not enough to induce the charge transfer. Reprinted figure with permission from [8]. Copyright (2009) by the American Physical Society

2.2 Electronic Structure

Early band structure calculations [9, 10] showed the presence of broad, half-filled antibonding bands of Cu-3d and O-2p states with $pd\sigma$ character across the Fermi level, implying the physical relevance of this hybridized state. The role of the other elements, such as La in La-214 systems or Y, Ba in Y-123 systems, in contributing to the electronic structure in high-T_c cuprates is believed to be minor due to the fact that the charge densities of these elements are mainly localized on their atomic sites, revealing an ionic character.

Cu is observed to exhibit variable valences, such as Cu^{1+} or Cu^{2+}, whereas Cu^{3+} is not common. Electrode potential measured in aqueous solution give a rough estimate of the relative stability of each oxidation state, since the cations coordinated by oxygen ions in aqueous solution resemble the crystalline environment [12]. Electrode potentials of the different Cu oxidations states are as follows [13].

$$
\begin{aligned}
&Cu^{2+} + e^- = Cu^+, \qquad && E^0 = 0.153\,\mathrm{eV} \\
&Cu^{3+} + e^- = Cu^{2+}, \qquad && E^0 = 2.4\,\mathrm{eV}
\end{aligned}
$$

The energy difference between Cu^{2+} and Cu^{1+} is quite low (0.153 eV), which may explain why Cu^{2+} can coexist with Cu^{1+} in the solid. However, the transition from Cu^{2+} to Cu^{3+} requires a large energy (2.4 eV), which makes the formation of Cu^{3+} chemically unfavorable. Indeed, X-ray absorption studies at the Cu-$L_{2,3}$ edge [14, 15] of $YBa_2Cu_3O_{7-\delta}$ showed the valence state of $Cu^{2+}(3d^9)$ at the Cu(2) atom in the CuO_2 plane, while the Cu^{3+} features are absent.

Based on this information, Fig. 2.6 illustrates the formation of the electronic structure based on Cu-3d and O-2p orbitals. In a cubic crystal field, the degenerate $3d^9$ states are split into e_g states with one hole and fully occupied t_{2g} states. Since Cu^{2+} is a common Jahn-Teller active ion, it leads to a further tetragonal distortion with the longest Cu-O bond out of plane, resulting in a further splitting of the partially occupied e_g states into a fully occupied $d_{3z^2-r^2}$ level and a $d_{x^2-y^2}$ level with one hole.

In the octahedral environment, the degenerate O-2p^6 levels are split into σ (along the planar Cu-O bond direction), $\pi_{//}$ (in-plane and perpendicular to the Cu-O bond

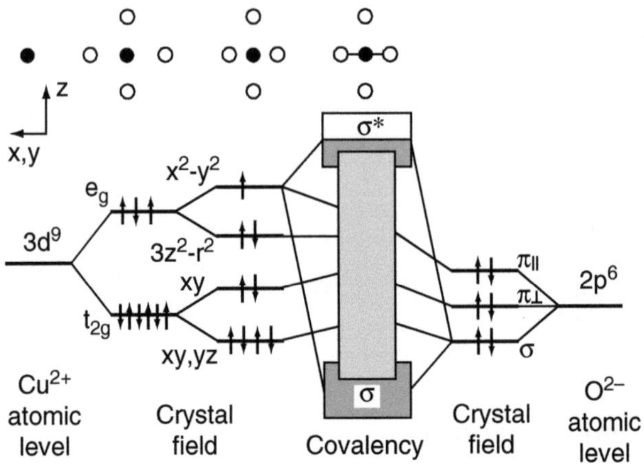

Fig. 2.6 Schematic picture of the electronic structure of the $pd\sigma$-hybridized Cu-3d and O-2p states. Reprinted figure with permission from [11]

direction) and $\pi_\perp$ (out-of-plane) orbitals. The strong covalency of the $d_{x^2-y^2}$ and p_σ orbitals gives rise to broad bonding (σ) and antibonding (σ^*) bands, whereas the orbital overlap between Cu-3d orbitals and O-2p_π orbitals is weak and results in narrow π-bands.

This single-electron band picture, however, results in a metallic ground state of the parent compound due to the odd number of electrons in the 3d-orbital. Particularly, band structure calculations with local density approximation (LDA) failed to predict the experimentally observed antiferromagnetic ground state with a large band gap ~1.5 eV [16, 17].

This apparent discrepancy is resolved by considering the Coulomb repulsion U on the Cu sites. This strong intra-atomic Coulomb repulsion originates from the spatially localized 3d-orbitals. The Hubbard model [18] is the minimal model that captures the insulating behavior observed in transition metal oxides. The Hamiltonian of the single-band Hubbard model reads

$$H = -t \sum_{ij,\sigma} \left(c^\dagger_{i,\sigma} c_{j,\sigma} + c_{i,\sigma} c^\dagger_{j,\sigma} \right) + U \sum_i n_{i,\uparrow} n_{i,\downarrow} \tag{2.1}$$

In the limit of $t \gg U$, this model simply reduces to the single-band tight-binding model with the bandwidth $2zt$, where z is the coordination number. The odd number of electrons places the chemical potential μ at the center of the band, resulting in a metallic ground state. The contribution of the on-site Coulomb repulsion is to impose an energy cost U for having two electrons at the same site. The role of U becomes more important as U/t increases. A large $U \gg t$ inhibits charge fluctuation of the type $d^n d^n \rightarrow d^{n+1} + d^{n-1}$, resulting in the splitting of the single band into the upper Hubbard band (UHB) and the Lower Hubbard band (LHB), thereby opening an

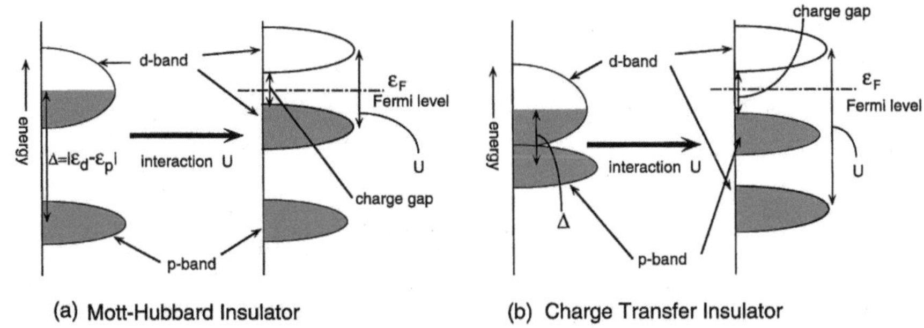

Fig. 2.7 Energy scheme of **a** the Mott-Hubbard insulator **b** the charge transfer insulator. Reprinted figure with permission from [20]. Copyright (1998) by the American Physical Society

energy gap in between. Even before the advent of high-T_c cuprates, it had been well established in $3d$-transition metal oxides that the characteristics of the $3d$ bands, i.e. the narrow bandwidth and the high intra-atomic electron Coulomb repulsion, require an explicit inclusion of the Hubbard U in the description of the electronic structure.

This is not the whole story. In $3d$ transition metal oxides, the charge degree of freedom involves O-$2p$ orbitals as well. This can be particularly important in high-T_c cuprates, where a strong $pd\sigma$-hybridization is expected from the band structure calculations. This introduces an additional energy scale Δ, which is called a charge transfer gap. This represents the energy required for charge fluctuations of the type $d^n d^n \rightarrow d^n \underline{L} + d^{n-1}$. Zaanen, Sawatzky and Allen [19] discussed the origin of the band gaps in transition metal compounds in relation to the relative strength of the Hubbard U and the charge transfer energy Δ. Considering the important role of the doped holes in cuprates, elucidating the nature of the band gap is particularly important in understanding their complex physics.

The schematic energy diagram of the transition metal oxides that takes into account both metal-$3d$ and O-$2p$ levels is depicted in Fig. 2.7. When the Hubbard U is smaller than the charge transfer energy Δ, it results in a dd-type band gap. On the other hand, when the Hubbard U is larger than the charge transfer energy Δ, the energy gap is of charge transfer type between the O-$2p$ and the metal-$3d$ bands. Photoemission studies [21, 22] measured the energy scales involved in this model for various copper oxides and showed that high-T_c cuprates are charge-transfer type insulators ($U \sim 6$–8 eV, $\Delta \sim 0.3$–2 eV, $U > \Delta$).

This is also supported by x-ray absorption studies at the Cu-$L_{2,3}$ edges [14, 15, 23]. The Cu valence in the CuO_2 planes does not change from Cu^{2+} across the entire doping range, whereas systematic changes are observed for the O-K edge. This indicates that the doped holes are mainly of oxygen character.

Thus, a general description of the correlated-electron physics on the CuO_2 plane requires an inclusion at the very least of the Cu-$3d_{x^2-y^2}$, O-$2p_x$ and O-$2p_y$ bands in the Hubbard model, as the one proposed by Emery [24].

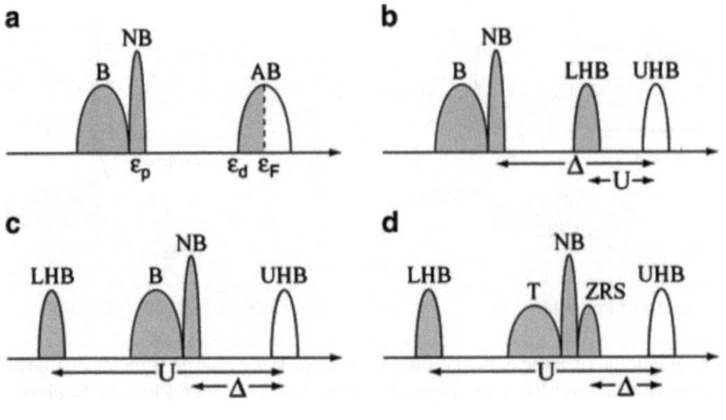

Fig. 2.8 Schematic energy diagram of the transition metal oxides with half-filling **a** without an electron-electron interaction **b** in a Mott-Hubbard insulator ($U < \Delta$) **c** in a charge transfer insulator ($U > \Delta$) **d** in a charge transfer insulator with Zhang-Rice singlets (case of high-T_c cuprates). AB refers to antibonding bands. NB refers to non-bonding bands. B refers to bonding bands. T refers to triplet hybridized states of the Cu-3d hole and O-2p hole. Reprinted figure with permission from [26]. Copyright (2003) by the American Physical Society

$$H = \sum_{ij\sigma} \epsilon_{ij} a^{\dagger}_{i\sigma} a_{j\sigma} + \frac{1}{2} \sum_{ij\sigma\sigma'} U_{ij} a^{\dagger}_{i\sigma} a_{i\sigma} a^{\dagger}_{j\sigma'} a_{j\sigma'}$$

Here, i is the lattice site of Cu or O atoms on the CuO_2 plane. $a^{\dagger}_{i\sigma}(a_{i\sigma})$ creates(destroys) a hole at the site i with the spin σ. The diagonal terms (ϵ_{ii}, U_{ii}) are the Cu-3d/O-2p band level ϵ_d, ϵ_p and the on-site Coulomb repulsion of the Cu-3d/O-2p orbitals U_{dd} and U_{pp}, respectively. The off-diagonal terms are the hopping integral between the Cu-3d and O-2p orbitals t_{dp} and the intersite Coulomb repulsion V_{dp}.

Zhang and Rice [25] showed that a proper downfolding of the Hamiltonian to remove the high energy states results in an effective one-band t-J model. The doped hole in this effective Hamiltonian is a singlet state of the symmetrized O-2p orbitals hybridized with the Cu-3d holes. This is called a Zhang-Rice singlet and it was proven to be energetically more stable than other mixtures of the Cu-3d and O-2p orbitals. This provides a framework to view all high-T_c cuprates as doped Mott insulators.

Figure 2.8 summarizes the discussion regarding the energy scheme of high-T_c cuprates presented above. In the absence of on-site Coulomb repulsion U, the ground state is a metallic state with a half-filled antibonding state of the Cu-3d and O-2p orbitals (Fig. 2.8a). Introducing the Hubbard U splits the top-lying antibonding band into the lower (LHB) and upper Hubbard bands (UHB). In the case of the Mott-Hubbard-type insulator ($U < \Delta$, Fig. 2.8b), the energy gap is of dd-type, whereas it is a charge transfer gap when $U > \Delta$ (Fig. 2.8c). The arguments by Zhang and Rice imply a further splitting of the non-bonding bands into triplet states and Zhang-Rice singlet (ZRS) states (Fig. 2.8d), where the doped holes first occupy the ZRS states.

2.3 Phase Diagram and Phenomenology

The phase diagram of the high temperature cuprate superconductors is depicted in Fig. 2.9. The undoped parent compound is an antiferromagnetic insulator. The antiferromagnetic ground state is rapidly suppressed by a small amount of hole doping $p_{min} \sim 0.05$ that slightly varies depending on the material. Above p_{min}, the long-range antiferromagnetic order is not detected anymore.

The physics of cuprates above p_{min} is characterized by diverse ground states that are sensitive to both doping and temperature. The doping level p_{opt} that shows the highest superconducting T_c is called optimal doping. The doping range below the optimal doping p_{opt} is called underdoped regime, whereas above the optimal doping, it is called overdoped regime. The normal state, i.e. temperatures above T_c in the absence of superconductivity, reveals a pseudogap state in the underdoped regime, whereas it shows a so-called strange metal phase in the overdoped regime, which is the least understood part of the phase diagram. The following sections will focus on explaining the various phases observed in the undoped and underdoped regimes.

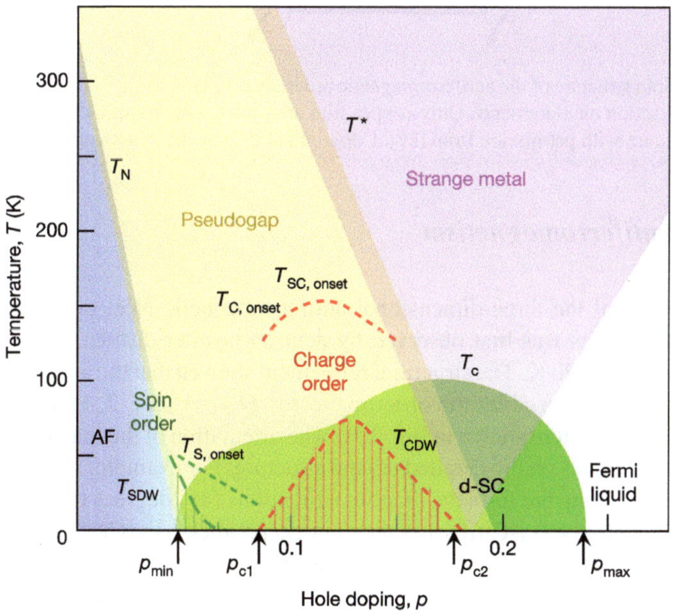

Fig. 2.9 Phase diagram of hole-doped high-temperature cuprate superconductors. Reprinted by permission from Nature [27]

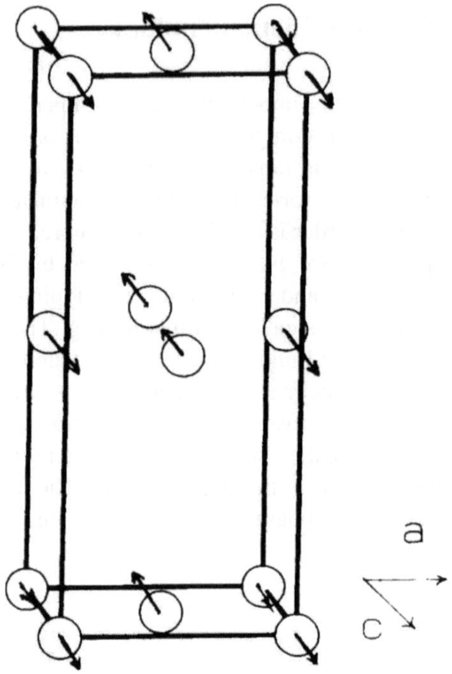

Fig. 2.10 Spin structure of the antiferromagnetic order observed in La_2CuO_4 proposed by neutron powder diffraction measurements. Only copper sites are shown with the spin directions as arrows. Reprinted figure with permission from [28]. Copyright (1987) by the American Physical Society

2.3.1 Antiferromagnetism

The existence of the three-dimensional antiferromagnetic Néel order in the high-T_c cuprate families was first observed by neutron powder diffraction on La_2CuO_4 [29] below $T_N = 220$ K. The structural refinement showed that the antiferromagnetic ordering is characterized by the ordering vector $\boldsymbol{Q} = (1/2, 1/2, 1)$, corresponding to the real space picture shown in Fig. 2.10. The orientation of the magnetic moments in the plane depends on the detailed crystal structure. For example, La_2CuO_4 has the Cu^{2+} moment along the in-plane diagonal directions (45° degrees to the Cu-O bond direction), whereas $YBa_2Cu_3O_6$ has the moment parallel to the Cu-O bond direction.

The size of the magnetic moment is measured to be 0.5–0.6 μ_B for the undoped cuprate compounds, much smaller than the classical spin magnetic moment of a free Cu^{2+} ion $g\langle S\rangle\mu_B \simeq 1.1\mu_B$ with $g \simeq 2.2$ and $\langle S\rangle = 1/2$. The large reduction of the magnetic moment is mainly explained by the quantum zero-point fluctuation of the spins [30], which stems from the fact that quantum mechanical spin vectors never, not even at zero temperature, lie along a certain axis, unlike classical spins. Since there is no exact solution of the two-dimensional antiferromagnetic Heisenberg model, numerous approximations have been carried out to study the ground states

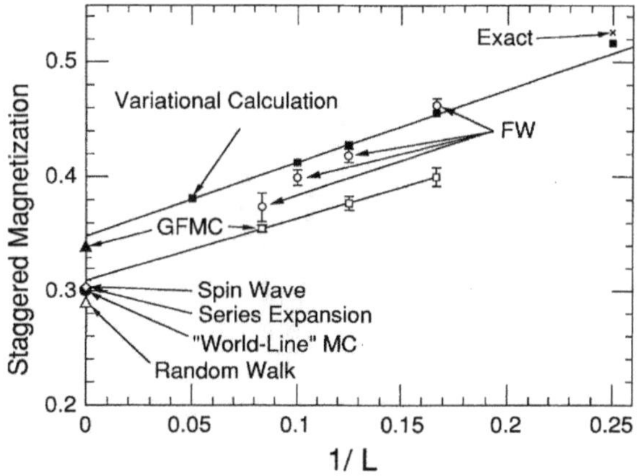

Fig. 2.11 Comparison of the various approximations to study the ground state of the two-dimensional antiferromagnetic Heisenberg model. The resulting staggered magnetization versus the system size was shown. Reprinted figure with permission from [30]. Copyright (1991) by the American Physical Society

of this system. As regards to the staggered moment, which is observed by neutron diffraction, consideration of the quantum fluctuation results in $\langle S \rangle \sim 0.3$ for the infinite system size ($L \to \infty$) as shown in Fig. 2.11, giving 60% of the classical moment $m \sim 0.66\mu_B$, slightly larger, but close to the experimental value.

A rough measurement of the exchange interaction J in the antiferromagnetic Heisenberg model can be performed by two-magnon Raman scattering, which gives $J \sim$ 100–150 meV [32–34]. A detailed description of the underlying magnetic interactions can be obtained by measurements of the dispersion of the magnetic excitations. Figure 2.12 shows the spin wave dispersion of La_2CuO_4 measured by inelastic neutron scattering [31]. The main features of the magnetic excitations are captured with the Heisenberg Hamiltonian with the nearest-neighbor interaction $J = 146.3 \pm 4$ meV and the cyclic interaction $J_C = 61 \pm 8$ meV. The second/third nearest-neighbor interaction were introduced to improve the fit quality, but the contribution is negligible ($J' \sim J'' \sim J_C/20$).

The origin of the antiferromagnetic Heisenberg interaction is explained by superexchange interaction, first proposed by Kramers [35], later refined by Anderson [36, 37]. The main driving force towards antiferromagnetism is that the electronic system comprising Cu^{2+} and neighboring oxygens gains an additional kinetic energy by antiparallel alignment of the neighboring Cu^{2+} spins.

Figure 2.13 illustrates how the hopping processes become different depending on the parallel/antiparallel alignment of the Cu^{2+} spins. With the antiparallel spin configuration, marked as red and green arrows, two different ways of consecutive hopping processes are allowed. When the $3d$-spins are parallel to each other, the first hopping of the electrons from the O-$2p$ orbitals is allowed, but the second hopping process is prohibited due to the Pauli principle. The fourth-order perturbation expansion [25, 39, 40] of the three-band Hubbard model results in a exchange interaction J as follows.

$$J = \frac{4t_{dp}^4}{(\Delta + V)^2} \left(\frac{1}{U_d} + \frac{2}{2\Delta + U_p} \right)$$

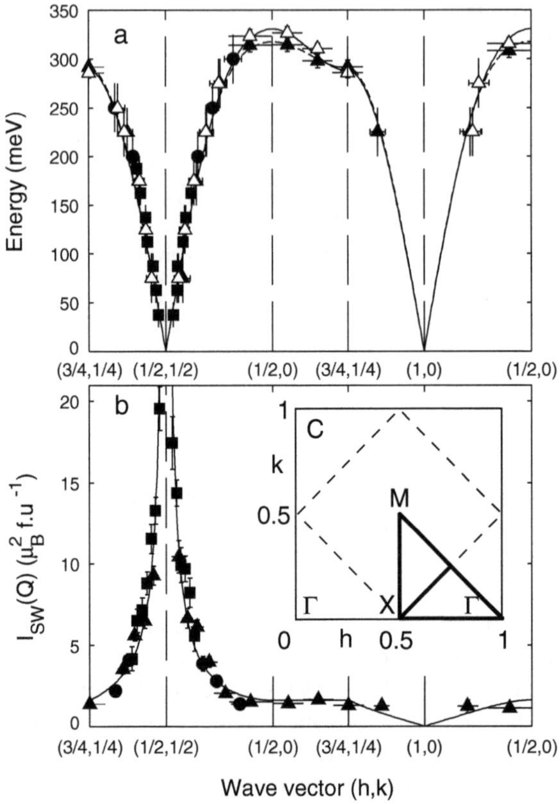

Fig. 2.12 **a** Spin waves of La_2CuO_4 along the high-symmetry directions in the 2D-Brillouin zone. Open (close) symbols were measured at T = 10 K (295 K). Squares are measured with the incident neutron energy $E_i = 250\,\text{meV}$, circles with $E_i = 600\,\text{meV}$ and triangles with $E_i = 750\,\text{meV}$. **b** Spin wave intensity in the 2D-Brillouin zone. Solid (dashed) line is a fit to the modelled spin-wave dispersion at 10 K (295 K) **c** 2D-Brillouin zone with the high-symmetry directions. Reprinted figure with permission from [31]. Copyright (2001) by the American Physical Society

Studies using Cu_2O_7 clusters to calculate J in various copper oxide compounds [41] have predicted $J \sim 0.15 - 0.16\,\text{eV}$, in reasonable agreement with the experimentally observed values.

Destruction of the antiferromagnetic ground state by a small amount of hole doping can be conceptually understood in the framework of doping a Mott insulator. This is illustrated in Fig. 2.14. A mobile hole introduced into the antiferromagnetic background gains a kinetic energy t upon hopping to a neighboring site. After the hopping, however, the Cu^{2+} spin is surrounded by a ferromagnetic environment with the energy cost $3/2J$, which is energetically unfavorable. In this conceptual picture, it is the relative strength of t compared to J, e.g. how mobile the hole is, that determines the stability of the antiferromagnetic order upon hole doping.

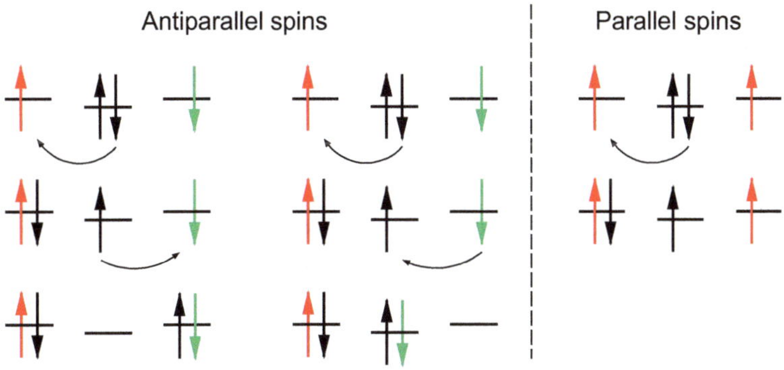

Fig. 2.13 Schematic picture of the superexchange mechanism. Red and green arrows represent the spins at the Cu-3d orbitals, whereas black arrows denote the spins at the O-2p orbitals. With the antiparallel Cu-3d spins, two ways of consecutive hopping are possible, whereas the second hopping is prohibited in the case of the parallel Cu-3d spin configuration. Figure adapted from [38]

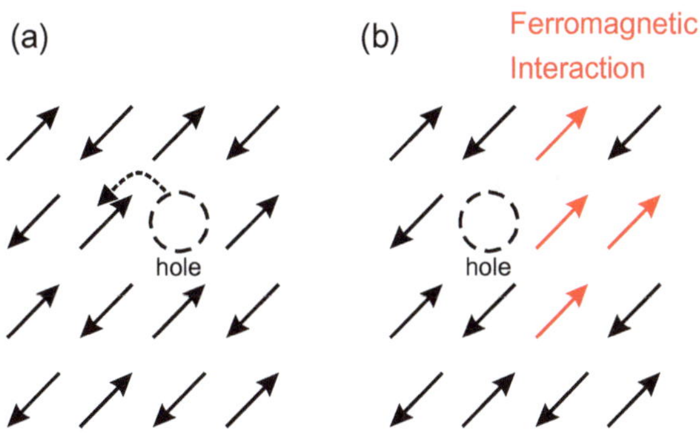

Fig. 2.14 Illustration of how a mobile hole is destructive to the antiferromagnetic order in a square-lattice Mott insulator **a** Presence of a hole in the antiferromagnetic background **b** Situation after the hopping of the hole. This induces an energy cost of three ferromagnetic interactions

If t is smaller than J, there is no energy gain in hopping and the induced hole is localized, leaving the antiferromagnetic background unaffected. However, if t is larger than J, the mobile hole jumps around the lattice sites, making the antiferromagnetic order unstable. Experimentally, only 0.03 holes/Cu are enough to perturb the antiferromagnetic order significantly, implying that the holes are highly mobile and the system finds another way to stabilize the system, e.g. density wave order or superconductivity.

2.3.2 *Superconductivity*

Superconductivity in high-T_c cuprates is, as its name suggests, characterized by its high superconducting transition temperature T_c that exceeds the theoretically proposed T_c limit of the BCS model of superconductivity (20–40 K) in the strong

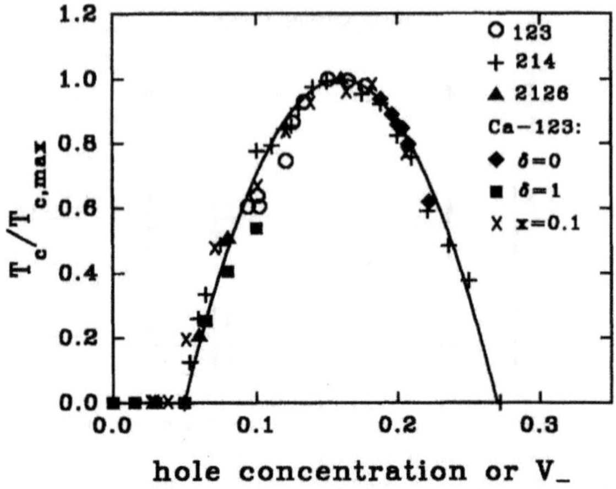

Fig. 2.15 T_c normalized by $T_{c,\max}$ versus hole concentration per Cu ion on the CuO_2 plane. Measurement data from different cuprate families are plotted together. Reprinted figure with permission from [43]. Copyright (1995) by the American Physical Society

coupling limit by McMillan [42]. The superconducting T_c of various high-T_c cuprates at their optimal dopings ranges from $T_c = 40\,\mathrm{K}$ in $La_{2-x}Sr_xCuO_4$ to $T_c = 134\,\mathrm{K}$ in $HgBa_2Ca_2Cu_3O_{8+x}$. The maximum superconducting T_c at ambient pressure was so far observed in the high-T_c cuprate families.

High-temperature superconductivity is achieved for an intermediate level of hole doping in the CuO_2 planes, after the complete suppression of the long-range antiferromagnetic order. It is commonly assumed that a universal superconducting dome exists in the phase diagram of high-T_c cuprates (Fig. 2.15), which is described by the empirical formula [44].

$$\frac{T_c}{T_{c,max}} = 1 - 82.6(p - p_c)^2 \tag{2.2}$$

p_c that gives the highest superconducting T_c is called an optimal doping. p_c is usually taken as 0.16 holes/Cu, which gives the minimum doping level required for superconductivity $p_{min} \sim 0.03$ and maximum doping $p_{max} \sim 0.27$. However, Karpinnen et al. pointed out that p_c is measured to be 0.12 in Bi-2212 families, which opens a possibility that $p_c = 0.16$ may not be universal, in which case the coupling between the intra-unit-cell CuO_2 layers may come into play [45].

The Ginzburg–Landau theory for superconductivity provides two characteristic length scales: the penetration depth λ, which is the length scale of the magnetic field penetration inside the superconductor, and the coherence length ξ, which represents the variation of the superconducting wave function in space.

High-temperature superconductivity in cuprates has a very short coherence length ξ and a long penetration depth λ, which according to Landau's criterion, make cuprate belong to the type II superconductors. A large penetration depth compared to a short coherence length energetically favors the formation of domain walls between superconducting and normal states above a lower critical field H_{c1}. This results in

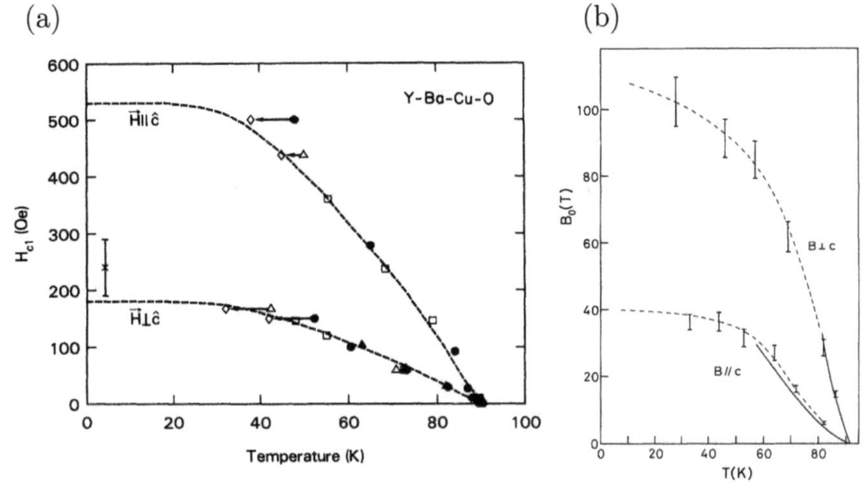

Fig. 2.16 **a** H_{c1} versus temperature. Reprinted figure with permission from [46]. Copyright (1989) by the American Physical Society. **b** H_{c2} versus temperature. Reprinted figure with permission from [47]. Copyright (1989) by the American Physical Society. Both data were measured in $YBa_2Cu_3O_{7-x}$ single crystals

the state of a superconducting vortex lattice until superconductivity is completely destroyed under magnetic fields above an upper critical field H_{c2}.

The coherence length can be obtained by the measurement of the upper critical field H_{c2} [48], using the formula

$$H_{c2}^{\perp} = \Phi_0/(2\pi\xi_{ab}^2)$$
$$H_{c2}^{//}/H_{c2}^{\perp} = \xi_{ab}/\xi_c$$

where Φ_0 is the magnetic flux quantum $\Phi_0 = 2.068 \times 10^{-15}\,\mathrm{T}\cdot\mathrm{m}^2$ and $\xi_{ab}(\xi_c)$ is the in-plane(out-of-plane) coherence length. Figure 2.16b shows the zero-temperature upper critical field of $H_{c2}^{//}(0) = 40\pm5\mathrm{T}$ and $H_{c2}^{\perp}(0) = 110 \pm 10\mathrm{T}$ [47]. This gives coherence lengths $\xi_{ab} = 17\,\text{Å}$ and $\xi_c = 6\,\text{Å}$.

The penetration depth can be measured by several experimental methods. The lower critical field H_{c1} is related to the penetration depth by

$$H_{c1} = \frac{\Phi_0}{4\pi\lambda_{ab}^2}\,[\ln(\lambda_{ab}/\xi_{ab}) + 0.5]$$

Measurement of H_{c1} in $YBa_2Cu_3O_{7-\delta}$ crystals were shown in Fig. 2.16b. This gives $H_{c1}^{//}(0) = 180\,\mathrm{Oe}$ and $H_{c1}^{\perp}(0) = 530\,\mathrm{Oe}$. Converted to the penetration depth, this implies that $\lambda_{ab}(0) = 1500\,\text{Å}$ and $\lambda_c(0) = 4500\,\text{Å}$. Microwave techniques [49] and μSR spin rotation measurements [50] on $YBa_2Cu_3O_{6.95}$ provided values consistent with this result.

Measurement of the Knight shift can distinguish different superconducting pairing scenarios, whether it is a singlet or triplet pairing. The Knight shift is change of the nuclear magnetic resonance (NMR) frequency arising from the interaction between the nuclear spin and paramagnetic electron spins. Singlet-pairing of superconductivity leads to a loss of the paramagnetic susceptibility that vanishes at T = 0 K, as was observed for BCS superconductors, whereas triplet-pairing is expected to show a remaining Knight shift at T = 0 K. The first Knight shift measurement was done for YBCO powders by Takigawa *et al.* [51]. They observed a decrease of the Knight shift below the superconducting transition that converges to zero after subtracting the chemical shift contribution, indicating singlet superconductivity in cuprates.

Unlike the isotropic *s*-wave gap in the BCS superconductors, high-T_c cuprates show an anisotropic *d*-wave gap in the superconducting state. The nature of the superconducting gap has a direct impact on the electronic contribution to the specific heat C_{el}, since it dramatically alters the density of states near the Fermi level. It is generally expected [54, 55] that C_{el} is described as $C_{el} = \alpha T^2$ in the presence of line nodes in the gap, where the coefficient α depends on the shape of the node. It also affects the specific heat of the mixed state under magnetic field below H_{c2}, where $C_{el}/T^2 = \beta B^{1/2} T$. The αT^2 and $\beta B^{1/2} T$ terms were observed for $YBa_2Cu_3O_{7-x}$ [56] and $La_{1.85}Sr_{0.15}CuO_4$ [57] in the limit of $T \gtrsim 0$ K, giving indirect evidence of the *d*-wave gap.

Most importantly, the magnitude of the *d*-wave superconducting gap was directly observed by angle-resolved photoemission spectroscopy (ARPES) experiments. Figure 2.17a shows the temperature-dependent ARPES spectra along $\Gamma - \bar{M}$ and $\Gamma - Y$ directions in the 2D-Brillouin zone (see the inset of the Fig. 2.17a). The spectral change at the superconducting transition is highly anisotropic. The spectra along the $\Gamma - \bar{M}$ direction, labelled as A in the figure, show a clear shift of the leading edge that reflects the opening of an energy gap. On the other hand, the spectra along the $\Gamma - Y$ direction, labelled as B in the figure, do not show a noticeable change at the superconducting transition. The position on the Fermi surface where the gap is maximum (zero) is called antinode (node).

The angular dependence of the superconducting gap is shown in Fig. 2.17b. The gap value becomes maximum as it approaches the antinodal region, whereas it is zero at the nodal region. It matches quite nicely with the *d*-wave functional form $\Delta(\boldsymbol{k}) = \Delta_0 \left[\cos(k_x a) - \cos(k_y a)\right]$ (solid line), demonstrating the *d*-wave character of the superconducting gap. The sign change of the superconducting gap function was later confirmed directly by Josephson tunneling experiments.

The mechanism of the high temperature superconductivity in cuprates, specifically the pairing glue of the superconductivity, is not clear yet. Several possibilities, such as resonating valence bond states [58], antiferromagnetic spin fluctuations [59], electron-phonon interaction [60], and charge fluctuations [61] have been extensively discussed, each of which pointed out important aspects that a complete theory of high-T_c superconductivity should possess. As pointed out in [62], a 'synergistic' pairing mechanism is needed that encompasses the salient features mentioned above.

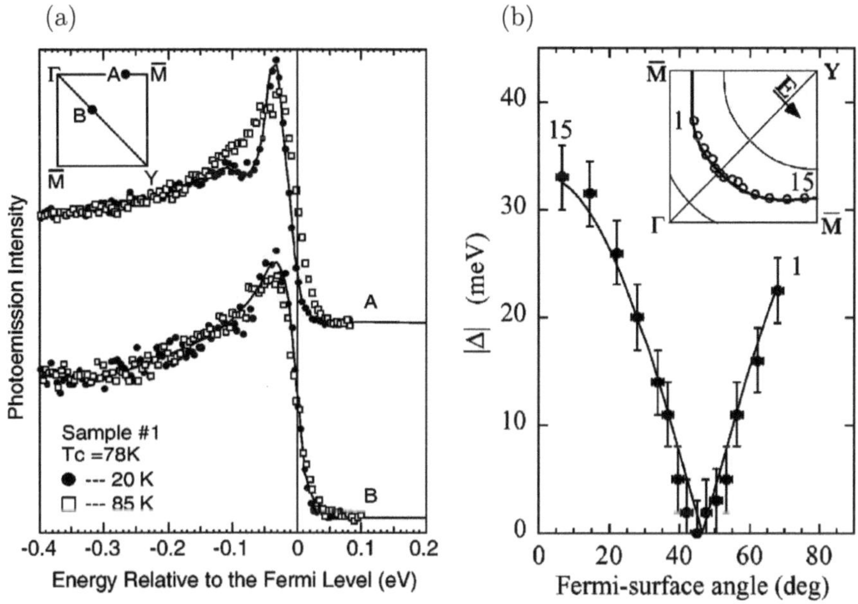

Fig. 2.17 **a** Temperature-dependent ARPES spectra measured above and below T_c from Bi2212. The spectra marked as A and B were measured at the points A and B in the 2D-Brilluoin zone drawn in the inset. Reprinted figure with permission from [52]. Copyright (1993) by the American Physical Society. **b** Superconducting gap measured on Bi2212 ($T_c = 87\,K$) at different angles on the Fermi surface. The data were taken at T = 10 K. Solid line represents a d-wave fit. Reprinted figure with permission from [53]. Copyright (1996) by the American Physical Society

2.3.3 Pseudogap

A gap-like feature above the superconducting transition temperature in cuprates was first observed by Warren *et al.* in $YBa_2Cu_3O_{6.7}$ using NMR [64]. The spin-lattice relaxation rate $1/T_1T$ of Cu(2) was observed to dramatically decrease below T = 100 K, well above the superconducting $T_c = 61\,K$. Since the dominant mechanism of spin-lattice relaxation is magnetic fluctuations, this result indicated a suppression of the low-energy spin excitations.

Further studies by Alloul *et al.* [65] showed a decrease of the ^{89}Y NMR Knight shift for superconducting YBCO powder samples well above T_c. The Knight shift is proportional to the electron-spin susceptibility and therefore, the result implies that the gradual decrease of the spin susceptibility is a universal feature present over a large doping range. Figure 2.18 shows the temperature dependence of the Knight shift measured in $YBa_2Cu_3O_{6.63}$ ($T_c = 62\,K$) and the spin susceptibility calculated from the data. The spin susceptibility continuously decreases by three times between T = 300 K and T = T_c, but still leaves a sizable value at T_c. The gradual decrease of the spin susceptibility over a large temperature range has been interpreted as an opening of an imperfect spin gap, a spin *pseudogap*.

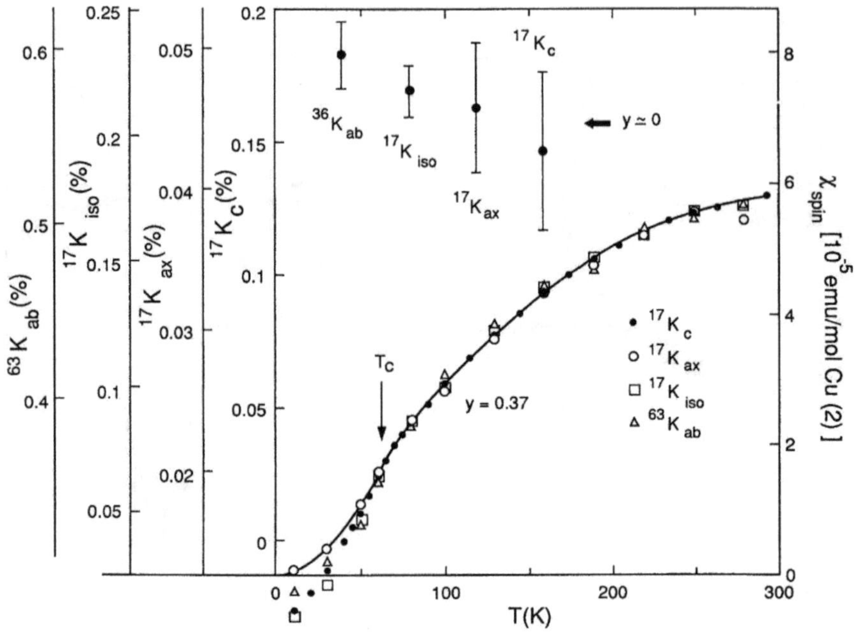

Fig. 2.18 Temperature dependence of the Knight shifts measured from various Cu and O components. The data were measured for $YBa_2Cu_3O_{6.63}$. Reprinted figure with permission from [63]. Copyright (1991) by the American Physical Society

Inelastic neutron scattering measures the imaginary part of the dynamic spin susceptibility. The emergence of a spin *pseudogap* above T_c was also observed in inelastic neutron scattering experiments. Figure 2.19 shows the imaginary part of the dynamic susceptibility $Im\left[\chi\left(\boldsymbol{q},\omega\right)\right]$ as a function of energy. At T = 5 K, below $T_c = 59$ K, a complete depression of the low energy spectral density (below 10 meV) is observed. Notably, at T = 75 K, even above T_c, the linear dispersion observed at high temperatures (T > 150 K) is not recovered, revealing a partial spin gap below $E_G = 16$ meV. The gap-like feature was not only observed in the spin channel, but also in the charge channel. Optical spectroscopic studies observed the formation of a *pseudogap* well above T_c in $YBa_2Cu_3O_{6.70}$ ($T_c = 63$ K), as described in Fig. 2.20a. Scanning tunneling microscopy (STM) studies on Bi-2212 also revealed a partial gap in the normal state of the cuprates. Figure 2.20b shows the scanning tunneling spectra of Bi-2212 as a function of temperature. At the lowest temperature T = 4.2 K, it shows a dip at zero bias and sharp peaks at $\pm\Delta_{gap}$, indicative of a superconducting gap. Above $T_c = 83$ K, the peaks at $\pm\Delta_{gap}$ disappear, but the the dip at zero bias remains throughout the measured temperature range above T_c.

ARPES experiments provide a direct observation of a pseudogap, i.e. a partial depletion of the spectral density on the Fermi surface. Several ARPES experiments on Bi-2212 [69, 70] revealed the momentum structure of the pseudogap in the normal

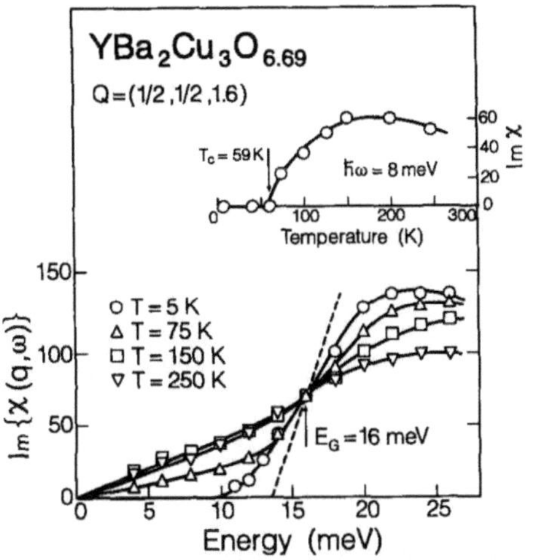

Fig. 2.19 Imaginary part of the spin susceptibility measured in $YBa_2Cu_3O_{6.69}$ at $\boldsymbol{Q} = (1/2, 1/2, 1.6)$ as a function of energy. Even at T = 75 K above T_c, a partial gap is present. Reprinted figure from [66]

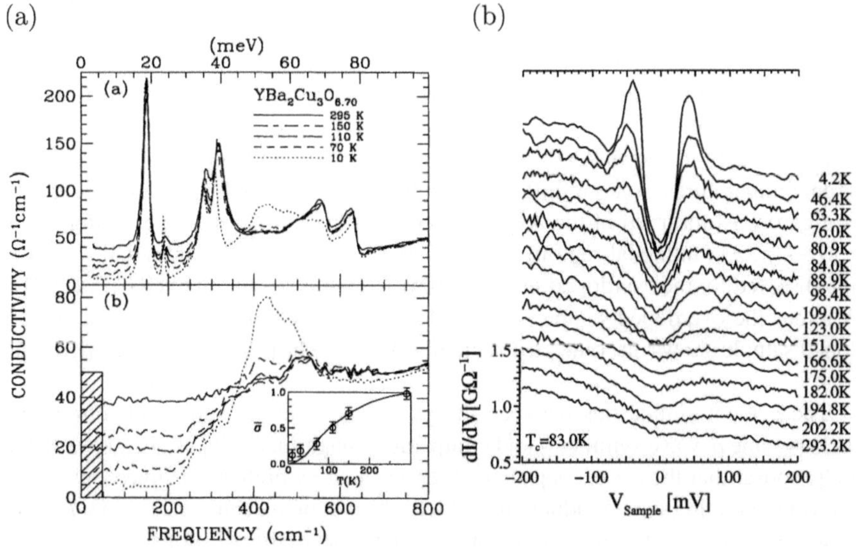

Fig. 2.20 **a** Optical conductivity spectra measured in $YBa_2Cu_3O_{6.67}$ (T_c = 63 K) at different temperatures. Reprinted figure with permission from [67]. Copyright (1993) by the American Physical Society. **b** Scanning tunneling spectra of the Bi-2212 as a function of temperature. The y-axis scale corresponds to the differential conductance measured at T = 295 K. The spectra are shifted by vertical offsets. Reprinted figure with permission from [68]. Copyright (1998) by the American Physical Society

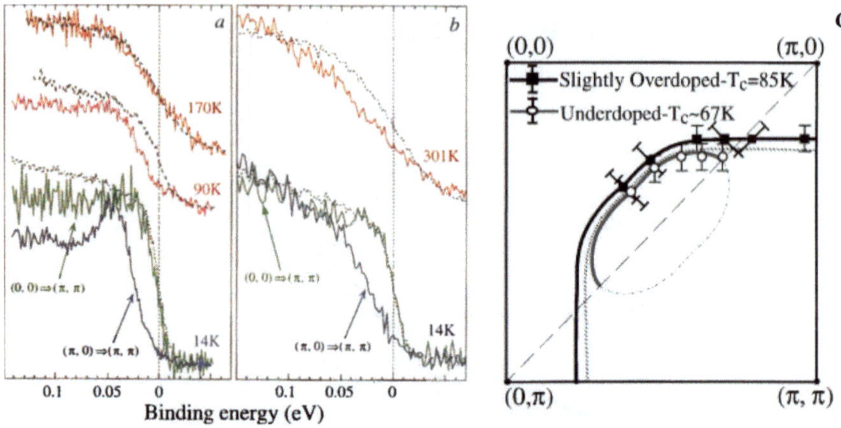

Fig. 2.21 Temperature-dependent APRES spectra of the underdoped Bi-2212 samples with **a** T_c = 83 K, **b** T_c = 10 K measured at the Fermi surface crossing along (0, 0)→(π, π)(nodal region) and (π,0)→(π, π)(antinodal region). Dotted lines are the reference Pt spectra. Reprinted by permission from Nature: [69]. **c** Fermi-level crossing measured from the Bi-2212 samples with two different hole dopings displayed in the 2D-Brillouin zone. The data were measured at T = 110 K, above T_c. Reprinted figure with permission from [70]. Copyright (1996) by the American Physical Society

state of the cuprates. In particular, they proved its existence in spectra measured in the antinodal region.

Figure 2.21a, b shows ARPES spectra measured for underdoped samples at different temperatures. A large shift of the leading edge near the Fermi level, indicative of a gap, was observed in the ARPES spectra in the Fermi surface crossing at the (π,0)→(π, π) (antinodal region) even above T_c. The gap in the antinodal region is not observed for the overdoped samples above T_c, which is summarized in Fig. 2.21c. Whereas the overdoped sample shows a fully connected Fermi surface, the underdoped sample shows disconnected segments, termed Fermi arcs, in the 2D-Brillouin zone.

The evident similarities between the pseudogap and the superconducting gap, such as same d-wave symmetry and comparable magnitude [69, 71, 72], have led to the proposal that the pseudogap state is represented by preformed superconducting pairs or fluctuating superconductivity, which achieves phase coherence below T_c [73]. However, experimental attempts to correlate the onset of the pseudogap known as T^* with the formation of the incoherent superconducting pairs has not been successful yet. Microwave measurements of the c-axis conductivity [74], Nernst effect measurements [75, 76], THz time-domain conductivity measurements [77], and optical spectroscopy [78] did observe signatures of superconducting fluctuations, but only in the vicinity of T_c, well below T^*.

The results of a detailed ARPES measurement with improved energy resolution (<10 meV) in Bi-2212 [80] suggested that the pseudogap is a distinct ground state that competes with superconductivity. They showed that the energy gap observed in

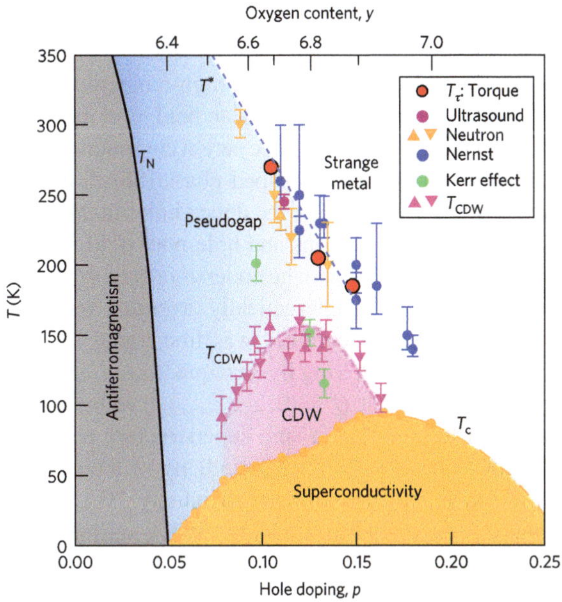

Fig. 2.22 Phase diagram of cuprates as a function of temperature and doping. The energy scales of the pseudogap T^* measured from different techniques mentioned in the text are marked in the phase diagram. Reprinted by permission from Nature Physics [79]

various dopings covering both underdoped and overdoped regimes is constant in the nodal region, whereas it decreases in the antinodal region as the hole doping increases. This nodal-antinodal dichotomy indicates that the two energy gaps are qualitatively different. The energy-momentum structure of the pseudogap in the ARPES spectra is clearly seen in the antinodal region inside the superconducting state, implying the existence of two distinct phases, not the pseudogap-superconducting gap transition. The back-bending feature of the antinodal energy dispersion, which implies a broken translational symmetry, was interpreted as evidence of a density wave order and was qualitatively reproduced to some degree with a mean field model using checkerboard-type charge order or short-range antiferromagnetic order [80, 81].

In light of the broken electronic symmetry assumed in the pseudogap state, several experimental evidences have suggested that the pseudogap is a state where the C_4 rotational symmetry is broken. Early electronic transport measurements [82] in underdoped YBCO pointed out that the in-plane resistivity anisotropy at low temperatures increases with decreasing hole doping in the pseudogap state, which is counterintuitive to the picture of the chain-assisted conductivity that increase as the hole doping increases. Various thermodynamic probes, such as inelastic neutron scattering [83], Nernst effect [84], torque magnetometry [79], and ultrasound [85], have observed an in-plane anisotropy present below the pseudogap energy scale T^*, which is summarized in Fig. 2.22. Further studies are necessary to elucidate the nature of the broken electronic symmetry in the pseudogap state.

2.3.4 Charge Density Waves

Right after the discovery of the high-temperature superconducting cuprates, it was theoretically proposed at the mean field level that a low concentration of holes introduced in the antiferromagnetic parent compound tend to self-localize to form charged domains, in particular a striped phase [86]. Studies using the t-J model also predicted phase separation of the doped holes in the antiferromagnetic background into hole-rich metallic regions and hole-poor antiferromagnetic domains [87, 88]. Phenomenologically, this can be understood as a segregation of the doped, non-magnetic holes to stabilize the energetically favorable antiferromagnetic domains.

Charge order in cuprates was first found in the 214 family of cuprates using elastic neutron scattering by Tranquada *et al.* [89]. Below T = 70 K, a charge order appears as a peak at $\boldsymbol{Q} = (0, 2\text{-}\delta_{charge}, 0)$ with $\delta_{charge} = 0.25$ in the reciprocal space. Upon further cooling, a spin superstructure peak is found at $\boldsymbol{Q} = (0.5, 0.5+\delta_{spin}, 0)$ with $\delta_{spin} = 0.125$. The periodicity of the spin order $1/\delta_{spin} = 8$ was measured to be twice the charge order periodicity $1/\delta_{charge} = 4$. This provides a real space picture depicted in Fig. 2.23d, where the antiferromagnetically ordered domains are separated by streams of non-magnetic hole-rich regions.

Further investigations on the stripe order revealed features qualitatively different from the mean-field predictions and the stripe order observed in other $3d$-transition metal oxides. For example, the appearance of the stripe order in $La_{1.875}Ba_{0.125}CuO_4$ does not open a gap and the system remains metallic [90], whereas the stripe order found in $La_{2-x}Sr_xNiO_4$ shows a sudden increase of the resistivity upon charge ordering [91], indicative of an opening of the charge gap. In addition, the low-temperature ground state of the moderately doped 214-cuprate shows superconductivity coexisting with the charge order, which was not expected from the mean-field theories.

The study of charge order is not only an interesting field on its own as charge ordering tendencies are widely observed in a variety of material systems, but also provides insights for the understanding of superconductivity by studying their relations to each other, i.e. whether charge order promotes or suppresses superconductivity. In cuprates, both charge order and superconductivity originate from the doped holes in the CuO_2 planes and therefore share the same electron density of states at the Fermi level. In this respect, one can expect that a subtle difference in energy between two ground states will lead to a complete suppression of one phase at zero temperature, but the reality from experiments shows phase coexistence, possibly nucleated by disorder.

One indication of the phase competition between charge order and superconductivity is the 1/8 anomaly observed in 214-cuprates. The 1/8 anomaly was first observed in $La_{2-x}Ba_xCuO_4$ compounds [92], as displayed in Fig. 2.24. A strong suppression of the superconducting T_c centered around $x = 1/8$ is observed, deviating from the assumed superconducting dome as a function of hole doping. Detailed doping-dependent studies of the stripe order in this compound [93] revealed that the strong suppression of superconductivity coincides with the emergence of the stripe order, as shown in Fig. 2.25.

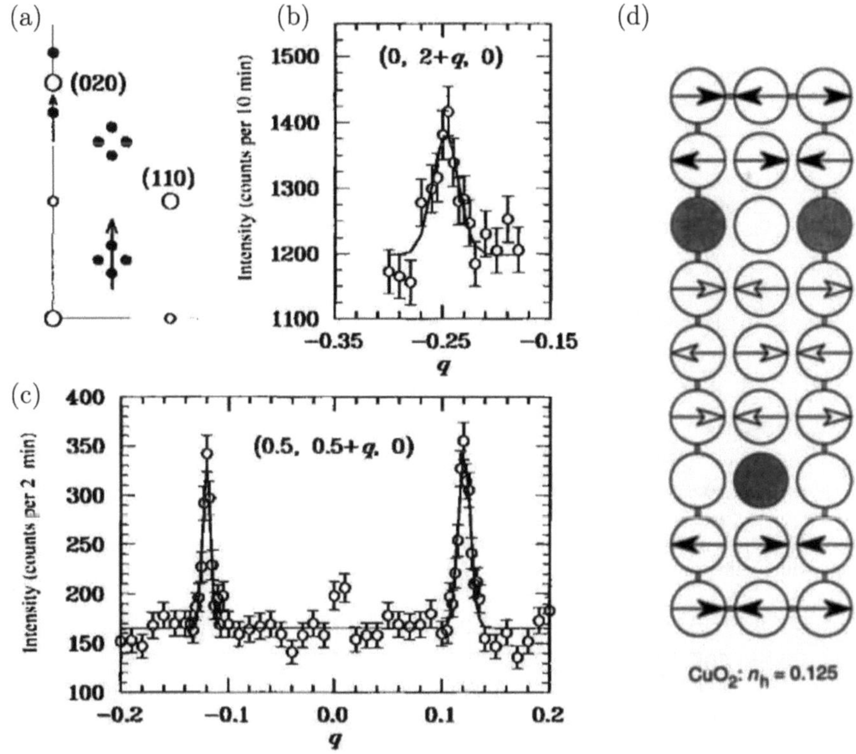

Fig. 2.23 **a** (H K 0) plane in the reciprocal space with the open circles representing the Bragg peaks and the solid circles denoting the charge and spin order superstructure peaks. Arrows are the scans done in the experiment. **b** Scan of the charge order superstructure peak along (0, 2+q, 0). **c** Scan of the spin order superstructure peak along (0.5, 0.5+q, 0). **d** Proposed real-space picture of the spin and charge stripe order for a hole density of $n_h = 1/4$. The circles represent the Cu atoms. The arrows represent the orientation of the magnetic moments in the Cu atoms. Holes are located at the atoms without arrows. Filled circles represent the presence of one hole. Reprinted by permission from Nature [89]

Early studies of the charge order in cuprates were confined to the 214-family of cuprates since experimental evidence of the charge order was only available from scattering experiments on this family of cuprates. STM can in principle detect a periodic charge modulation in the real space by mapping out the local density of states (LDOS) and calculating the Fourier transform of the LDOS map. Although it is an attractive tool to provide the real-space charge distribution as well as the spectroscopic information by scanning tuneling spectroscopy (STS), it is only sensitive to the sample surface and therefore the interpretation is often criticized by the argument that it may not reflect the bulk properties. Nevertheless, the same cuprate physics may govern the phenomenology near the surface and one expects to obtain a general idea about the cuprate physics also from STM studies.

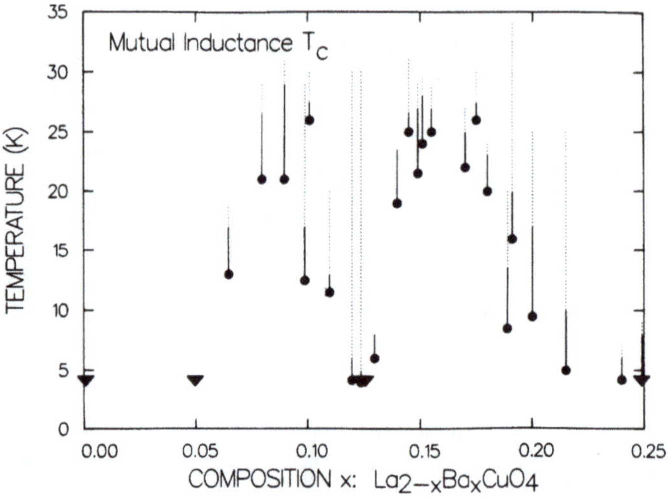

Fig. 2.24 Superconducting T_c versus the Ba substitution ratio x in $La_{2-x}Ba_xCuO_4$. Reprinted figure with permission from [92]. Copyright (1988) by the American Physical Society

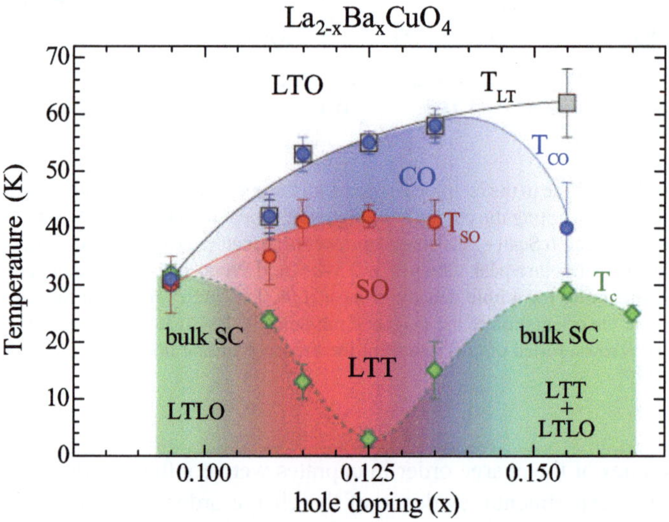

Fig. 2.25 Phase diagram of $La_{2-x}Ba_xCuO_4$ in the range $x = 0.095$–0.155. T_{LT} is a low-temperature LTO-LTT structural transition temperature. T_{CO}(T_{SO}) is a charge(spin) order transition temperature. T_c is a superconducting transition temperature. Reprinted figure with permission from [93]. Copyright (2011) by the American Physical Society

STM studies were mainly performed for the $Bi_2Sr_2CaCu_2O_{8+\delta}$ and $Na_xCa_{2-x}CuO_2Cl_2$ families, where an *in-situ* cleaving of the surface is possible. There are some STM studies available for 214-compounds and YBCO, but atomic resolution was not achieved, which is related to the issue of cleaving and surface degradation. The first signature of a periodic charge modulation in STM studies was obtained by Hoffman *et al.* in BSCCO. So-called quasi-particle interference (QPI), which is a generalized Friedel oscillation caused by impurities, was observed with dispsersing in energy, as suggested by Capriotti *et al.* [94].

This finding, however, imposed a stringent constraint on the interpretation of the STM data on charge order, since doped cuprates are in general dirty and all the STS data should possess substantial QPI signals, which should be distinguished from the possible charge order signal. The unambigious observation of charge order by STM was first presented by Vershinin *et al.* [95] in the pseudogap state of BSCCO. They observed a non-dispersive peak in the Fourier transform of the LDOS below 40 meV, which is not consistent with QPI. The center of the peak gives a periodicity of $4.7a_0$ along the Cu-O bond direction. Extensive studies by STM encompass the discussion of the charge order characteristics, such as stripe/checkerboard nature [96, 97], bond-centered charge modulation [97], d-wave form factor [98], and vestigial nematic nature of the charge order [99], which are under intense debate. The detailed discussion on the STM studies in cuprates is beyond the scope of this thesis and can be found in reviews [100, 101].

2.3.4.1 Charge Density Waves in $YBa_2Cu_3O_{6+x}$

The charge density waves in $YBa_2Cu_3O_{6+x}$ were first discovered using resonant inelastic x-ray scattering (RIXS) by Ghiringhelli *et al.* [102]. Energy-resolved scans along the Cu-O bond direction, as shown in Fig. 2.26, observed a narrow quasi-elastic peak with an incommensurate wavevector of ~0.31 reciprocal lattice units

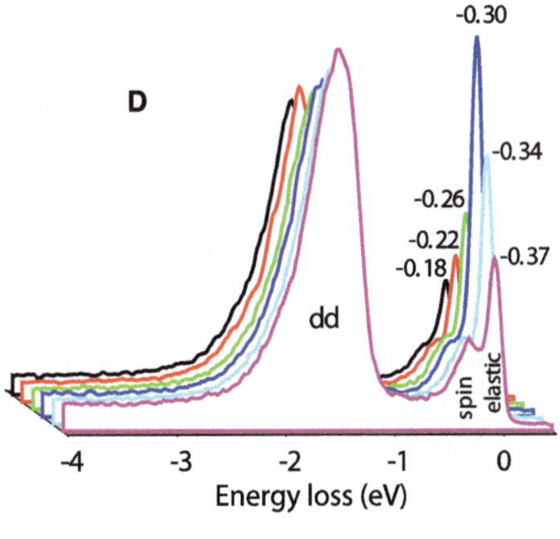

Fig. 2.26 Momentum dependence of the RIXS spectra measured at the $Cu - L_3$ edge for $Nd_{1.2}Ba_{1.8}Cu_3O_7$. From [102]. Reprinted with permission from AAAS

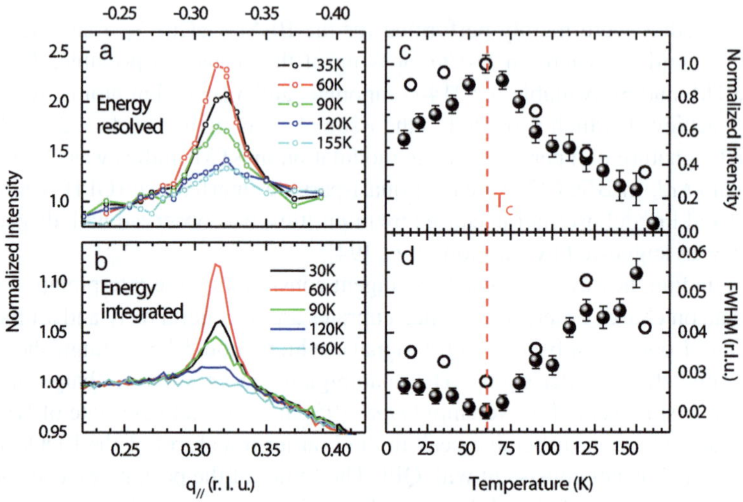

Fig. 2.27 **a** Momentum dependence of the CDW quasi-elastic intensity as a function of temperature **b** Momentum dependence of the energy-integrated CDW signal as a function of temperature. **c** Normalized intensity of the CDW intensities as a function of temperature. **d** FWHM of the CDW peak as a function of temperature. Open circles are from the energy-resolved scans and solid circles are from the energy-integrated measurements. From [102]. Reprinted with permission from AAAS

that occurs below T = 150 K, indicating the existence of charge correlations with the periodicity of ∼3.2 lattice constants. The intensity and the width of the peak shows a strong temperature dependence, increasing in intensity and correlation length below T = 150 K and decreasing below T = T_c, as displayed in Fig. 2.27. The resonance of the charge order signal at the absorption edge of the Cu(2) site and the absence of the signal at the Cu(1) site energy indicate that the charge order arises from the Cu atoms in the CuO_2 planes.

Charge density waves in $YBa_2Cu_3O_{6+x}$ have been extensively studied thanks to its low chemical disorder with defects confined in the CuO chain layer and relatively clean CuO_2 planes. Below are the list of the properties that were observed for the CDW in YBCO by x-ray scattering experiments. The qualitative features mentioned below are commonly observed in other family of cuprates as well.

(i) Observed in the underdoped regime
CDW is observed in underdoped YBCO, in the doping range of $p = 0.08 - 0.16$, as shown in Fig. 2.28a. The onset temperature of the CDW depends on the doping and ranges from T = 100 K to T = 150 K above the superconducting dome, inside the pseudogap regime. It shows the highest transition temperature at the doping level of $p \sim 0.12$, analogous to the 1/8 anomaly observed in 214-cuprates.

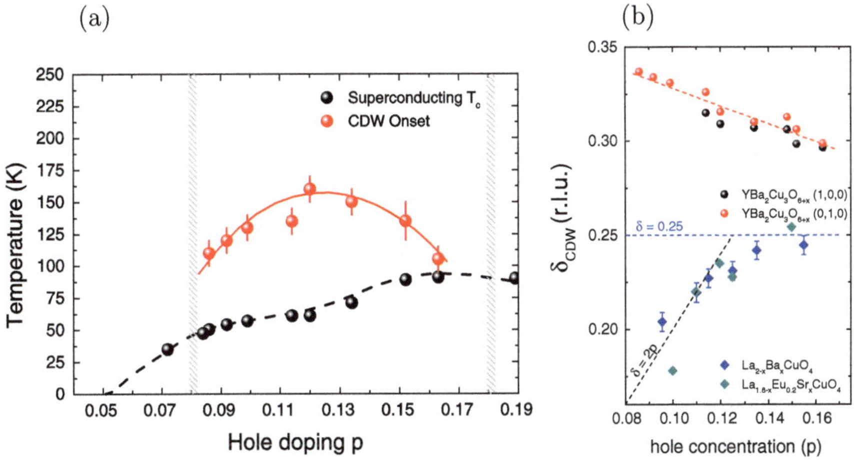

Fig. 2.28 **a** Doping dependence of the CDW onset temperature and the superconducting T_c. **b** Doping dependence of the CDW incommensurability δ_a (black circles) and δ_b (red circles) of the underdoped $YBa_2Cu_3O_{6+x}$. The results from the 214-cuprates are also displayed. Reprinted figure with permission from [103]. Copyright (2014) by the American Physical Society

(ii) Bi-directional and incommensurate
The CDW quasi-elastic peaks are observed both along a- and b-axis, at the wave vector $\boldsymbol{Q} = (\delta_a, 0), (0, \delta_b)$ in the 2D-Brillouin zone, with the incommensurate values of δ_a, δ_b. Figure 2.28b shows the doping dependence of the CDW incommensurability δ_a and δ_b. The incommensurability shows a linear dependence that decreases upon hole doping. δ_a and δ_b differ by 0.01 r.l.u., possibly related to the small orthorhombicity in underdoped YBCO.

(iii) Quasi two-dimensional nature
The CDW domains are anisotropic with moderate in-plane correlation lengths and very small out-of-plane correlation length, showing a quasi two-dimensional nature. The correlation length can be calculated as the inverse of the width of the quasi-elastic CDW peak. The in-plane longitudinal correlation lengths measured at $T = T_c$ are 20–70 Å depending on the hole doping level, which gives 5–18 unit cells, whereas the out-of-plane correlation length is less than one lattice constant. The momentum scan along L (out-of-plane direction) shows that the CDW intensity is centered at half-integer values [104]. This observation is similar to the charge stripes observed in 214-cuprates, which was explained by the presence of long-range Coulomb repulsion between the holes in the neighboring layers that induces an inter-layer anti-correlation [105].

(iv) Absence of static incommensurate magnetic order
CDW in YBCO is observed without any signature of static magnetic order. This is in stark contrast to the stripe order in 214-cuprates, where static charge and magnetic order coexist in the form of non-magnetic hole-rich regions serving as domain boundaries of the hole-poor antiferromagnetically ordered domains.

Static incommensurate magnetic order was observed in YBCO, but only for low doping levels ($p \leq 0.085$) [106–108]. This barely overlaps with the doping regime where CDW is observed ($0.084 \leq p \leq 0.163$) [103], which might imply a phase competition between CDW and static incommensurate magnetic order.

(v) Competition with superconductivity
At all doping levels where it is present, the 2D charge order appears before the onset of supercondcutivity. The 2D charge order domains grow upon cooling below its onset temperature, until the temperature reaches T_c. Below the superconducting T_c, the CDW loses intensity and the correlation length decreases. This was interpreted as sign of the competition between the superconductivity and the charge order.
Another direct evidence of the competition between superconductivity and CDW was given by x-ray scattering experiments under high magnetic field. The magnetic field suppresses superconductivity and in turn, it is expected to enhance the CDW if the two phases compete with each other. Figure 2.29 illustrates the change of the CDW under magnetic field as a function of temperature. The effect of the magnetic field is seen as an increase of the CDW intensity and correlation length below T_c, whereas the CDW is barely affected above T_c. Above the critical field of this doping ($\sim$15 T), where superconducting long-range order is expected to be completely suppressed, the enhancement of the CDW saturates, showing the classic behavior of an order parameter. This provides convincing evidence that CDW and superconductivity are competing phases.

(vi) Strong electron-phonon coupling
Due to the presence in cuprates of a substantial electron-phonon interaction, an anomaly in the phonon spectra near the CDW ordering vector is expected. This was recently investigated by inelastic x-ray scattering experiments on underdoped YBCO [109, 110]. These studies observed a substantial broadening of the low-energy phonon modes at the superconducting transition near the

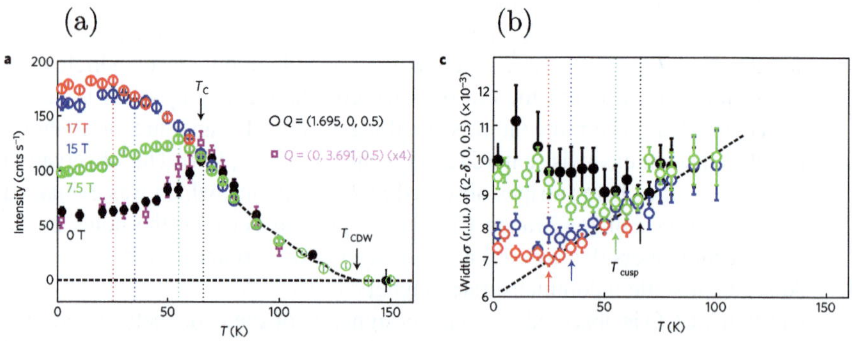

Fig. 2.29 **a** CDW intensity **b** CDW peak width under different strengths of magnetic field as a function of temperature. Reprinted by permission from Nature Physics [104]

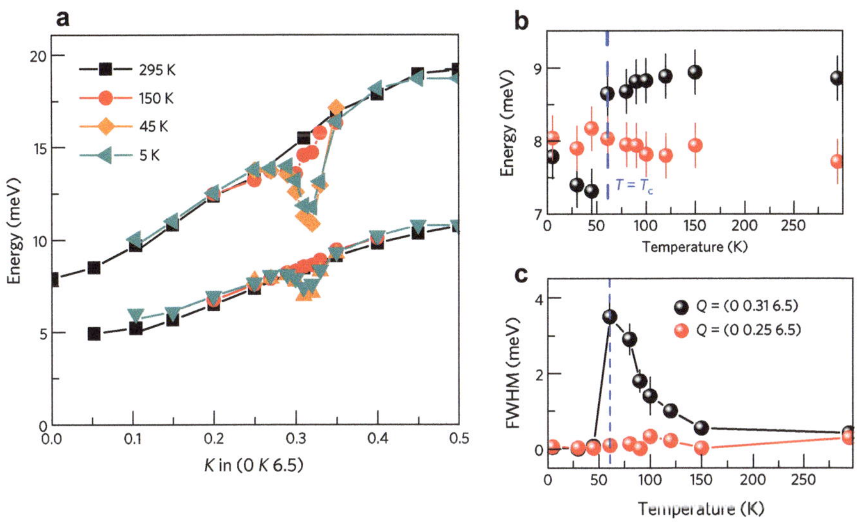

Fig. 2.30 **a** Dispersion of the low energy phonons in $YBa_2Cu_3O_{6.67}$ measured at various temperatures. Temperature dependence of the acoustic **b** phonon energy and **c** FWHM as a function of temperature. Reprinted by permission from Nature Physics [109]

CDW wave vector. In addition, Le Tacon *et al.* [109] observed a giant phonon softening of a low-energy phonon mode near the CDW wave vector below T_c. Surprisingly, the effect of the softening is strongest at the superconducting T_c, and not at the CDW onset temperature $T_{\text{onset}} = 150$ K, implying the complex intertwined nature of the two phases (Fig. 2.30).
Phonon anomalies at the CDW wave vector were subsequently observed in other families of cuprates, in $La_{1.875}Ba_{0.125}CuO_4$ [111], $Bi_2Sr_2CaCu_2O_{8+\delta}$ [112], as expected in light of the strong electron-phonon coupling universally present in cuprates.

Recently, Gerber *et al.* [114] observed the emergence of a three-dimensional CDW in $YBa_2Cu_3O_{6.67}$ ($p = 0.12$, $T_c = 67$ K) by carrying out x-ray scattering experiments under high magnetic field. Above magnetic fields exceeding $H \sim 15$ T, the elastic intensity with a strong out-of-plane correlation appears at the integer L on top of the 2D-CDW signal centered at the half-integer L. The 3D-CDW was observed to have the same incommensurability as the 2D-CDW. The correlation volume of the 3D-CDW measured at B ~ 30 T ($\xi_b^{3D} \sim 180$ Å, $\xi_c^{3D} \sim 50$ Å) is almost two orders of magnitude larger than that of the 2D-CDW ($\xi_b^{2D} \sim 40$ Å, $\xi_c^{2D} \sim 7$ Å), revealing its long-range nature. Following studies [115, 116] revealed that the 3D-CDW is unidirectional, present only along the *b*-axis, and is observed in similar doping range as the 2D-CDW.

The observed 3D-CDW under high magnetic field seems to be closely linked to the thermodynamic transition observed in the high field measurements of the Hall coefficient [117, 118], NMR [119] and sound velocity measurements [113, 120].

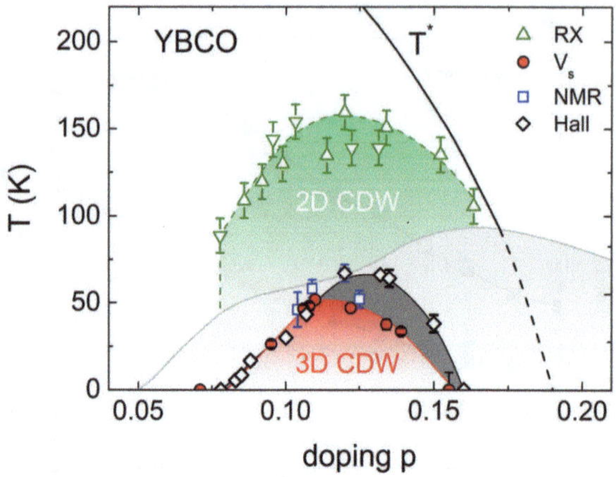

Fig. 2.31 Phase diagram of the charge order in YBCO. The onset temperatures of the 2D-CDW measured by resonant x-ray scattering and the thermodynamic transition temperatures under high field were displayed. Reprinted by permission from npj Quantum Materials [113]

The doping dependence of the onset temperatures measured by these methods in fact match well with each other, as shown in Fig. 2.31.

The thermodynamic ground state under high field in the underdoped regime is characterized by small nodal electron pockets revealed by quantum oscillation experiments [121]. This is qualitatively different from the large hole-like Fermi surface observed in the overdoped regime, which is well-reproduced by band structure calculations.

The origin of the electron pockets in the underdoped regime is still under intense debate. Density wave orders are proposed to induce such an electron pocket, but the explanation is not yet satisfactory. The existence of the bidirectional 2D charge order was suggested to produce an electron pocket by band-folding, but there is a suspicion that the 2D CDW might be too short-ranged to produce such a large change in the Fermi surface [122]. The NMR data [119] and the long-range nature measured by x-ray scattering [114–116] suggests that the 3D-CDW might be responsible for the Fermi surface reconstruction. However, it was pointed out that the unidirectional nature of the 3D-CDW cannot produce a closed pocket. Further investigations are needed to elucidate the origin of the Fermi surface reconstruction in the normal state of the underdoped regime in cuprates.

The main driving force of the CDW formation in YBCO is also not yet clear. In the case of Bi-2201, the Fermi surface nesting scenario, borrowed from the conventional Peierls systems, was brought up to explain the charge order. Combined studies of RXS, ARPES and STM [123] argued that the nesting vector connecting two hot spots of the neighboring Fermi arcs, and thus namely a Fermi-arc instability, drives the charge order in Bi-2201. However, YBCO does not provide a clean, bulk-representing surface for clean ARPES measurement and therefore, a reliable measurement of its Fermi surface is lacking so far. Nevertheless, a linear decrease of the CDW incommensurability upon increasing hole doping is in line with the Fermi-arc nesting scenario. The length of the Fermi arc increases as the hole doping increases, which results in decreasing length of the nesting vector and incommensurability.

2.3.4.2 Charge Density Waves in Bi- and Hg-Based Cuprates

The successful measurement of charge order using Resonant x-ray scattering (RXS) triggered extensive investigations in other cuprate families. Resonant x-ray scattering on La-doped Bi-2201 [123] ($0.11 < p < 0.14$) and Bi-2212 [124] ($0.07 < p < 0.13$) revealed the existence of incommensurate charge density waves. The observed CDW in Bi-based cuprates shares several features observed with the one observed in YBCO: (i) The RXS intensity of the CDW appears in the underdoped regime below the onset temperatures of ∼200 K above the superconducting dome. (ii) The measured incommensurability lies in the range 0.24–0.3 r.l.u, comparable to the observed incommensurability 0.3–0.35 r.l.u. in YBCO. (iii) The temperature scans show a competition with superconductivity, but it is rather weak compared to YBCO.

For La-doped Bi-2201 systems, combined studies with ARPES showed that the observed incommensurability is much larger than the nesting vector that connects the antinodal points on the Fermi surface. Instead, they found a link between the incommensurability and the nesting vector connecting the ends of the neighboring Fermi arcs, so-called hot-spots, which suggests an intimate relation between pseudogap and CDW in the underdoped regime.

The CDW was finally observed in Hg-based cuprates as well, establishing the charge ordering tendency as a ubiquitous property of high-T_c cuprates. RXS experiments [125] on Hg-1201 compounds ($T_c = 72$ K) observed charge density waves with incommensurate wave vector $\boldsymbol{Q} \sim 0.28$ r.l.u. below the onset temperature ∼200 K. However, in Hg-based cuprates, the temperature dependence of the RXS intensity shows a negligible change at the superconducting transition, which might reflect mesoscopic coexistence between both states.

References

1. Bednorz JG, Müller KA (1986) Zeitschrift für Physik B Con- densed Matter 64:189–193
2. Barišić N et al (2013) Proc Natl Acad Sci 110:12235–12240. https://doi.org/10.1073/pnas.1301989110. https://www.pnas.org/content/110/30/12235.full.pdf
3. Jorgensen JD et al (1990) Phys Rev B 41:1863–1877. https://doi.org/10.1103/PhysRevB.41.1863. Accessed 4 Feb 1990
4. Cava RJ et al (1990) Phys C: Supercond 165:419–433
5. Calestani G, Rizzoli C (1987) Nature 328:606–607
6. Andersen N et al (1999) Phys C: Supercond 317–318:259–269. https://doi.org/10.1016/S0921-4534(99)00066-0
7. Liang R, Bonn D, Hardy WN (2000) Phys C: Supercond 336:57–62. https://doi.org/10.1016/S0921-4534(00)00091-5
8. Alloul H et al (2009) Rev Mod Phys 81, RMP:45–108
9. Mattheiss LF (1987) Phys Rev Lett 58, PRL:1028–1030

10. Massidda S et al (1987) Phys Lett A 122:198–202
11. Fink J et al (1989) IBM J Res Dev 33:372–381
12. Geballe TH, Koster G (2006) Handbook of high-temperature superconductivity, Chap 8: What Tc can teach about superconductivity. Springer, Berlin, p 327
13. Cotton FA et al (1999) Advanced inorganic chemistry, Chap 17: The elements of the first transition series. A Wiley- Interscience Publication, p 855
14. Bianconi A et al (1987) Solid State Commun 63:1009–1013. https://doi.org/10.1016/0038-1098(87)90650-8
15. Tranquada JM et al (1987) Phys Rev B 35:7187–7190. https://doi.org/10.1103/PhysRevB.35.7187. Accessed 13 May 1987
16. Uchida S et al (1991) Phys Rev B 43:7942–7954. https://doi.org/10.1103/PhysRevB.43.7942. Accessed 10 April 1991
17. Orenstein J et al (1990) Phys Rev B 42:6342–6362. https://doi.org/10.1103/PhysRevB.42.6342. Accessed 10 Oct 1990
18. Hubbard J, Flowers BH (1963) Proc R Soc Lond Ser A Math Phys Sci 276:238–257. https://doi.org/10.1098/rspa.1963.0204. https://royalsocietypublishing.org/doi/pdf/10.1098/rspa.1963.0204
19. Zaanen J, Sawatzky GA, Allen JW (1985) Phys Rev Lett 55:418–421. https://doi.org/10.1103/PhysRevLett.55.418. Accessed 4 July 1985
20. Imada M, Fujimori A, Tokura Y (1998) Rev Mod Phys 70:1039–1263. https://doi.org/10.1103/RevModPhys.70.1039. Accessed 4 Oct 1998
21. Shen Z-X et al (1987) Phys Rev B 36:8414–8428. https://doi.org/10.1103/PhysRevB.36.8414. Accessed 16 Dec 1987
22. Fujimori A et al (1989) Phys Rev B 39:2255–2260. https://doi.org/10.1103/PhysRevB.39.2255. Accessed 4 Feb 1989
23. Merz M et al (1998) Phys Rev Lett 80:5192–5195. https://doi.org/10.1103/PhysRevLett.80.5192. Accessed 23 June 1998
24. Emery VJ (1987) Phys Rev Lett 58, PRL:2794–2797
25. Zhang FC, Rice TM (1988) Phys Rev B 37:3759–3761. https://doi.org/10.1103/PhysRevB.37.3759. Accessed 7 Mar 1988
26. Damascelli A, Hussain Z, Shen Z-X (2003) Rev Mod Phys 75:473–541. https://doi.org/10.1103/RevModPhys.75.473. Accessed 2 April 2003
27. Keimer B et al (2015) Nature 518:179–186
28. Vaknin D et al (1987) Phys Rev Lett 58:2802–2805. https://doi.org/10.1103/PhysRevLett.58.2802. Accessed 26 June 1987
29. Shirane G et al (1987) Phys Rev Lett 59, PRL:1613–1616
30. Manousakis E (1991) Rev Mod Phys 63:1–62. https://doi.org/10.1103/RevModPhys.63.1. Accessed 1 Jan 1991
31. Coldea R et al (2001) Phys Rev Lett 86:5377–5380. https://doi.org/10.1103/PhysRevLett.86.5377. Accessed 23 June 2001
32. Lyons KB et al (1988) Phys Rev Lett 60:732–735. https://doi.org/10.1103/PhysRevLett.60.732. Accessed 8 Feb 1988
33. Lyons KB et al (1988) Phys Rev B 37:2353–2356. https://doi.org/10.1103/PhysRevB.37.2353. Accessed 4 Feb 1988
34. Blumberg G et al (1997) Science 278:1427–1432. https://doi.org/10.1126/science.278.5342.1427. https://science.sciencemag.org/content/278/5342/1427.full.pdf
35. Kramers H (1934) Physica 1:182–192. https://doi.org/10.1016/S0031-8914(34)90023-9
36. Anderson PW (1950) Phys Rev 79:350–356. https://doi.org/10.1103/PhysRev.79.350. Accessed 2 July 1950
37. Anderson PW (1959) Phys Rev 115, PR:2–13
38. Koch E, cond-mat: Chap 7 Exchange mechanism. https://www.cond-mat.de/events/correl12/manuscripts/koch.pdf
39. Ogata M, Shiba H (1988) J Phys Soc Jpn 57:3074–3088. https://doi.org/10.1143/JPSJ.57.3074

40. Emery VJ, Reiter G (1988) Phys Rev B 38:4547–4556. https://doi.org/10.1103/PhysRevB.38.4547. Accessed 7 Sept 1998
41. Mizuno Y, Tohyama T, Maekawa S (1998) Phys Rev B 58:R14713–R14716. https://doi.org/10.1103/PhysRevB.58.R14713. Accessed 22 Dec 1998
42. McMillan WL (1968) Phys Rev 167:331–344. https://doi.org/10.1103/PhysRev.167.331. Accessed 2 March 1968
43. Tallon JL et al (1995) Phys Rev B 51:12911–12914. https://doi.org/10.1103/PhysRevB.51.12911. Accessed 18 May 1995
44. Presland M et al (1991) Phys C: Supercond 176:95–105. https://doi.org/10.1016/0921-4534(91)90700-9
45. Karppinen M et al (2003) Phys Rev B 67:134522. https://doi.org/10.1103/PhysRevB.67.134522. Accessed 13 April 2003
46. Krusin-Elbaum L et al (1989) Phys Rev B 39, PRB:2936–2939
47. Nakao K et al (1989) Phys Rev Lett 63, PRL:97–100
48. Chaudhari P et al (1987) Phys Rev B 36, PRB:8903–8906
49. Hardy WN et al (1993) Phys Rev Lett 70:3999–4002. https://doi.org/10.1103/PhysRevLett.70.3999. Accessed 25 June 1993
50. Bonn DA et al (1993) Phys Rev B 47:11314–11328. https://doi.org/10.1103/PhysRevB.47.11314. Accessed 17 May 1993
51. Takigawa M et al (1989) Phys Rev B 39:7371–7374. https://doi.org/10.1103/PhysRevB.39.7371. Accessed 10 April 1989
52. Shen Z-X et al (1993) Phys Rev Lett 70:1553–1556. https://doi.org/10.1103/PhysRevLett.70.1553. Accessed 10 March 1993
53. Ding H et al (1996) Phys Rev B 54:R9678–R9681. https://doi.org/10.1103/PhysRevB.54.R9678. Accessed 14 Oct 1996
54. J Exp Theor Phys Lett 64:690–694 (1996)
55. J Exp Theor Phys Lett 65:491–496 (1997)
56. Wright DA et al (1999) Phys Rev Lett 82:1550–1553. https://doi.org/10.1103/PhysRevLett.82.1550. Accessed 7 Feb 1999
57. Fisher RA et al (2000) Phys Rev B 61:1473–1476. https://doi.org/10.1103/PhysRevB.61.1473. Accessed 2 Jan 2000
58. Anderson PW (1987) Science 235:1196–1198. https://doi.org/10.1126/science.235.4793.1196. https://science.sciencemag.org/content/235/4793/1196.full.pdf
59. Schrieffer JR, Wen X-G, Zhang S-C (1988) Phys Rev Lett 60:944–947. https://doi.org/10.1103/PhysRevLett.60.944. Accessed 10 March 1988
60. Kulić ML (2000) Phys Rep 338:1–264. https://doi.org/10.1016/S0370-1573(00)00008-9
61. Imada M (2005) Phys Rev B 72:075113. https://doi.org/10.1103/PhysRevB.72.075113. Accessed 7 Aug 2005
62. Plakida N (2010) High-temperature cuprate superconductors, experiments, theory and applications. Springer, Berlin, p 478
63. Takigawa M et al (1991) Phys Rev B 43:247–257. https://doi.org/10.1103/PhysRevB.43.247. Accessed 1 Jan 1991
64. Warren WW et al (1989) Phys Rev Lett 62:1193–1196. https://doi.org/10.1103/PhysRevLett.62.1193. Accessed 10 March 1989
65. Alloul H, Ohno T, Mendels P (1989) Phys Rev Lett 63:1700–1703. https://doi.org/10.1103/PhysRevLett.63.1700. Accessed 16 Oct 1989
66. Rossat-Mignod J et al (1991) Phys B: Condensed Matter 169:58–65
67. Homes CC et al (1993) Phys Rev Lett 71, PRL:1645–1648
68. Renner C et al (1998) Phys Rev Lett 80:149–152. https://doi.org/10.1103/PhysRevLett.80.149. Accessed 1 Jan 1998
69. Ding H et al (1996) Nature 382:51–54
70. Marshall DS et al (1996) Phys Rev Lett 76, PRL:4841–4844
71. Norman MR et al (1998) Nature 392:157–160
72. Loeser AG et al (1996) Science 273:325–329

73. Emery VJ, Kivelson SA (1995) Nature 374:434–437
74. Grbić MS et al (2009) Phys Rev B 80, PRB:094511
75. Xu ZA et al (2000) Nature 406:486–488
76. Wang Y, Li L, Ong NP (2006) Phys Rev B 73:024510. https://doi.org/10.1103/PhysRevB.73.024510. Accessed 2 Jan 2006
77. Bilbro LS et al (2011) Nat Phys 7:298–302
78. Corson J et al (1999) Nature 398:221–223
79. Sato Y et al (2017) Nat Phys 13:1074
80. Hashimoto M et al (2014) Nat Phys 10:483–49
81. He R-H et al (2011) Science 331:1579–1583
82. Ando Y et al (2002) Phys Rev Lett 88, PRL:137005
83. Fauqué B et al (2006) Phys Rev Lett 96, PRL:197001 (2006)
84. Daou R et al (2010) Nature 463:519
85. Shekhter A et al (2013) Nature 498:75
86. Zaanen J, Gunnarsson O (1989) Phys Rev B 40, PRB:7391–7394
87. Kivelson SA, Emery VJ, Lin HQ (1990) Phys Rev B 42:6523–6530. https://doi.org/10.1103/PhysRevB.42.6523. Accessed 10 Oct 1990
88. Emery VJ, Kivelson SA, Lin HQ (1990) Phys Rev Lett 64:475–478. https://doi.org/10.1103/PhysRevLett.64.475. Accessed 4 Jan 1990
89. Tranquada JM et al (1995) Nature 375
90. Tranquada JM et al (2008) Phys Rev B 78, PRB:174529
91. Cheong S-W et al (1994) Phys Rev B 49:7088–7091. https://doi.org/10.1103/PhysRevB.49.7088. Accessed 10 March 1994
92. Moodenbaugh AR et al (1988) Phys Rev B 38:4596–4600. https://doi.org/10.1103/PhysRevB.38.4596. Accessed 7 Sept 1988
93. Hücker M et al (2011) Phys Rev B 83:104506. https://doi.org/10.1103/PhysRevB.83.104506. Accessed 10 March 2011
94. Capriotti L, Scalapino DJ, Sedgewick RD (2003) Phys Rev B 68:014508. https://doi.org/10.1103/PhysRevB.68.014508. Accessed 1 July 2003
95. Vershinin M et al (2004) Science 303:1995–1998
96. Wise D et al (2008) Nat Phys 4:696–699
97. Kohsaka Y et al (2007) Science 315:1380–1385. https://doi.org/10.1126/science.1138584. https://science.sciencemag.org/content/315/5817/1380.full.pdf
98. Hamidian MH et al (2015) Nat Phys 12:150
99. Mukhopadhyay S et al (2019) Proc Natl Acad Sci 116:13249–13254
100. Fischer Ø et al (2007) Rev Mod Phys 79:353–419. https://doi.org/10.1103/RevModPhys.79.353
101. Yazdani A, da Silva Neto EH, Aynajian P (2016) Ann Rev Condensed Matter Phys 7:11–33. https://doi.org/10.1146/annurev-conmatphys-031214-014529
102. Ghiringhelli G et al (2012) Science 337:821–825. https://doi.org/10.1126/science.1223532. https://science.sciencemag.org/content/337/6096/821.full.pdf
103. Blanco-Canosa S et al (2014) Phys Rev B 90, PRB:054513
104. Chang J et al (2012) Nat Phys 8:871
105. Mv Zimmermann et al (1998) Europhys Lett 41:629–634. https://doi.org/10.1209/epl/i1998-00204-2
106. Hinkov V et al (2008) Science 319:597–600
107. Haug D et al (2009) Phys Rev Lett 103:017001. https://doi.org/10.1103/PhysRevLett.103.017001. Accessed 1 June 2009
108. Haug D et al (2010) New J Phys 12:105006
109. Le Tacon M et al (2014) Nat Phys 10:52–58
110. Blackburn E et al (2013) Phys Rev B 88:054506. https://doi.org/10.1103/PhysRevB.88.054506. Accessed 5 Aug 2013
111. Miao H et al (2018) Phys Rev X 8:011008. https://doi.org/10.1103/PhysRevX.8.011008. Accessed 1 Jan 2018

112. Chaix L et al (2017) Nat Phys 13:952–956
113. Laliberté F et al (2018) npj Quantum Mater 3:11
114. Gerber S et al (2015) Science 350:949–952. https://doi.org/10.1126/science.aac6257. https://science.sciencemag.org/content/350/6263/949.full.pdf
115. Chang J et al (2016) Nat Commun 7:11494. https://doi.org/10.1038/ncomms11494
116. Jang H et al (2016) Proc Natl Acad Sci 113:14645–14650. https://doi.org/10.1073/pnas.1612849113. https://www.pnas.org/content/113/51/14645.full.pdf
117. LeBoeuf D et al (2007) Nature 450:533–536
118. LeBoeuf D et al (2011) Phys Rev B 83:054506. https://doi.org/10.1103/PhysRevB.83.054506. Accessed 5 Feb 2011
119. Wu T et al (2011) Nature 477, 816MC Times Cited:366 Cited References Count:30, 191–194
120. LeBoeuf D et al (2013) Nat Phys 9:79–83
121. Doiron-Leyraud N et al (2007) Nature 447:565–568
122. Gannot Y, Ramshaw BJ, Kivelson SA (2019) Phys Rev B 100:045128. https://doi.org/10.1103/PhysRevB.100.045128. Accessed 4 July 2019
123. Comin R et al (2014) Science 343:390–392. https://doi.org/10.1126/science.1242996. https://science.sciencemag.org/content/343/6169/390.full.pdf
124. da Silva Neto EH et al (2014) Science 343:393–396
125. Tabis W et al (2014) Nat Commun 5:5875

Chapter 3
Experimental Methods

3.1 Synchrotron Radiation

The two fundamental reasons for using x-ray synchrotron radiation in research is to take advantage of its two main properties: (i) x-ray energy tunability, which allows for different type of experiments and in particular to exploit resonances, and (ii) high brilliance that far exceeds that of the conventional x-ray tubes. Brilliance is a useful parameter that quantitatively compares the quality of different radiation sources. It takes into account not only the flux density, but also the degree of the beam collimation and its monochromaticity, as defined below.

$$\text{Brilliance} = \frac{\text{photons/second}}{(\text{mrad})^2(\text{mm}^2\ \text{source area})(0.1\%\ \text{bandwidth})}$$

This is well described in Fig. 3.1. Traditional x-ray tubes used in laboratories exhibit a brilliance of the order of $10^8 - 10^9$ photons/s/mm^2/mrad^2/0.1%BW. A typical synchrotron x-ray beam has a brilliance of several orders of magnitude higher than x-ray tubes, ranging from $10^{12} - 10^{21}$ photons/sec/mm^2/mrad^2/0.1%BW. Highly brilliant x-rays provide huge benefits in research. A synchrotron beam not only saves a significant amount of measurement time thanks to the high flux, but also allows one to study otherwise inaccessible experimental features with low scattering cross sections, such as inelastic scattering. In addition, experiments with a small sample volume become possible owing to its high flux density.

Figure 3.2 shows a schematic picture of the x-ray radiation in the synchrotron. In third-generation synchrotron facilities, electrons move around a vacuum storage ring with a relativistic speed $v \lesssim c$, of several GeV in energy. The circumference of the storage ring ranges from several hundreds to thousands meters. Inside the ring, electron bunches are accelerated and their trajectory is bent when passing through a bending magnet or an insertion device, such as a wiggler or an undulator, as a result of which the x-rays are generated. These x-rays are further processed via several

H. Kim, *Uniaxial Pressure Study of Charge Density Waves in a High-T_c Cuprate Superconductor*, Springer Theses,
https://doi.org/10.1007/978-3-030-99898-1_3

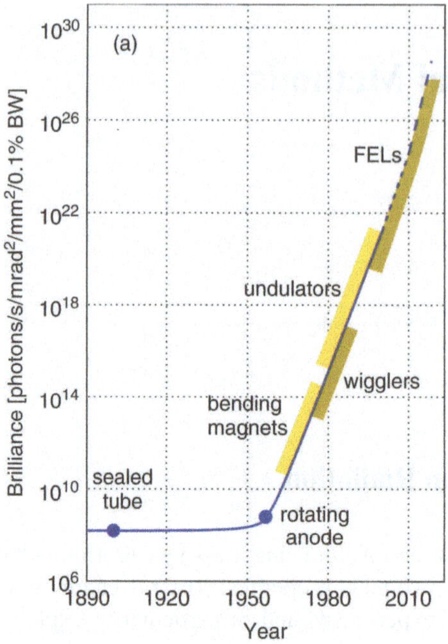

Fig. 3.1 The historical development of the different x-ray sources and their brilliances. Republished with permission of John Wiley & Sons—Books, from [1]

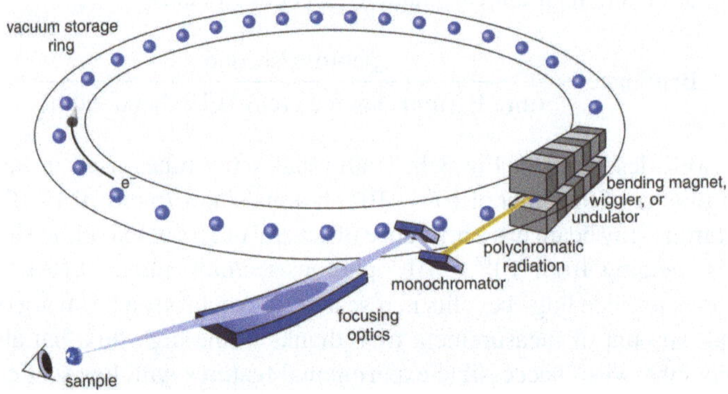

Fig. 3.2 Schematic picture of how x-ray is produced in the synchrotron radiation source. Republished with permission of John Wiley & Sons—Books, from [1]

x-ray optics (e.g. monochromator, focusing optics, etc.) for the individual purpose of each experimental station at the end of so-called beamlines.

Figure 3.3 provides a simplified, but intuitive picture of how relativistic electrons can emit electromagnetic radiation with the wavelength of x-rays. An oscillating charge at a relativistic speed produces compressed wavefronts of electromagnetic radiation with an extremely short wavelength along its moving direction. The strong

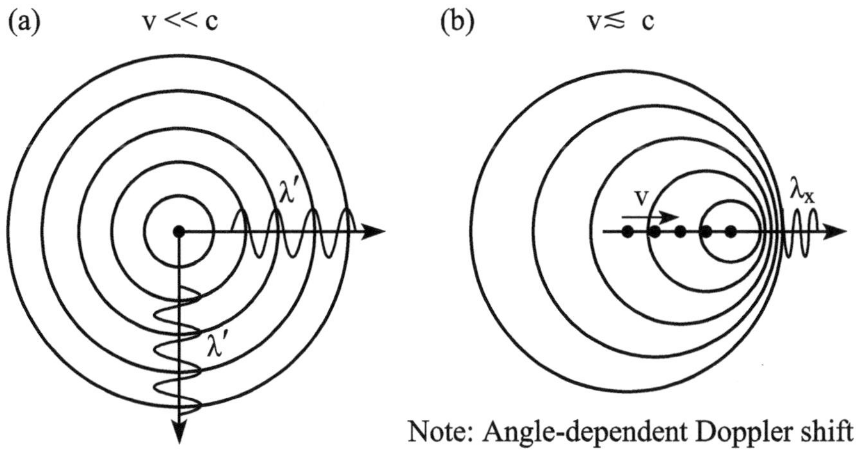

Fig. 3.3 Electromagnetic radiation from oscillating charges **a** at non-relativistic speed of electrons $v \ll c$. **b** at relativistic speed of electrons $v \lesssim c$. Circles represent the wavefronts of the emitted electromagnetic radiation by oscillating charges. Reproduced with permission of Cambridge University Press through PLSclear, from [2]

angular dependence of the radiation wavelength provides a very narrow light 'cone' along the direction of motion, i.e. x-rays with a very small emission angle. This is in stark contrast to the case of a non-relativistic oscillating charge that emits radiation with a nearly constant wavelength at all angles.

Bending magnets and insertion devices such as undulators are the essential building elements of third-generation synchrotrons that cause an acceleration of relativistic electrons. As shown in Fig. 3.4a, bending magnets apply a uniform magnetic field on the electron beam path, which deflects the electron motion and induces a fan of radiation. The emission angle of the bending magnet radiation is $\sim 1/\gamma$, where γ is the Lorentz factor $\gamma = \frac{E}{m_0 c^2}$ of the electron ($m_0 c^2$ is the rest mass energy of the electron).

$\theta \sim 1/\gamma$ is taken as a natural opening angle of the storage ring. As mentioned previously, typical third-generation synchrotrons have a storage-ring energy E of several GeV. Using the formula $\gamma = 1957E[\text{GeV}]$, γ gives a value of 2000–16000, which results in a emission angle of 0.5–0.06 mrad.

The bending magnet radiation is a 'white' beam that covers a wide spectral range. The critical energy of the bending magnet radiation, defined as the photon energy that divides the total emitted power in half, is determined by the storage ring energy and the magnet strength, $\hbar\omega_c\,[\text{keV}] = 0.665E^2[\text{GeV}]\,B[T]$. This is taken as a representative photon energy of the bending magnet with a sufficient flux. For a low storage ring energy of 2 GeV and a permanent magnet strength of 1 T, the critical energy is $\hbar\omega = 2.66\,\text{keV}$, which lies in the tender or intermediate x-ray range ($\sim$1–5 keV). This implies that the generation of hard x-rays using bending magnets is realized in facilities with higher storage ring energy and/or with the aid of an increased magnet strength by employing superconducting magnets.

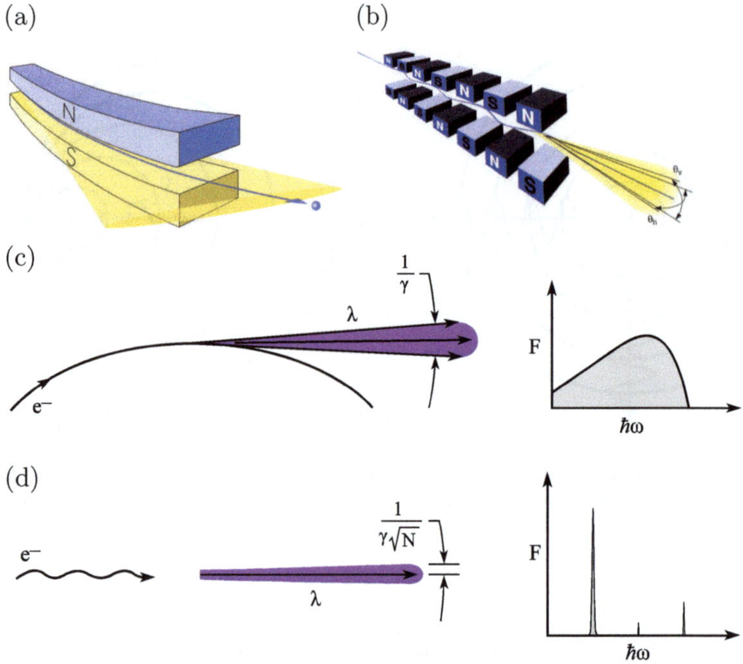

Fig. 3.4 **a** Schematic picture of a bending magnet radiation. **b** Schematic picture of an undulator radiation. **a** and **b** are reproduced with permission of John Wiley & Sons—Books, from [1]. **c** Characteristics of the bending magnet radiation. It has an emission angle of $\sim 1/\gamma$ that covers a wide range of energy. **d** Characteristics of the undulator radiation. It has an emission angle of $\sim 1/(\gamma\sqrt{N})$ and the harmonics of a selected energy calculated by the undulator equation. **c** and **d** are reproduced with permission of Cambridge University Press through PLSclear, from [2]

Insertion devices are placed between the bending magnet segments in the storage ring. An example of an insertion device is an undulator, which is a series of very small alternating bending magnets as shown in Fig. 3.4b. Undulator radiation is characterized by a smaller emission angle $\theta \sim 1/(\gamma\sqrt{N})$ (Fig. 3.4d) compared to that of a bending magnet. Moreover, an undulator produces specific harmonics with photon energy calculated by the undulator equation

$$\lambda = \frac{\lambda_u}{2\gamma^2}\left(1 + \frac{K^2}{2} + \gamma^2\theta^2\right)$$

where λ is the resulting wavelength of the undulator radiation, λ_u is the undulator period, γ is the Lorentz factor, K is a dimensionless parameter with the value $K = 0.934\lambda_u$ [cm] $B_0[T]$, θ is the angular spread of the undulator radiation. Since the term in parentheses $1 + \frac{K^2}{2} + \gamma^2\theta^2$ is of the order of 1, the wavelength of the undulator radiation is mainly determined by the undulator period λ_u and the Lorentz factor γ of the storage ring. A typical γ of $10^3 - 10^4$ implies that an undulator period of several mm to cm is required to reach the different x-ray energy ranges.

3.2 X-Ray Scattering Technique

X-ray scattering is a photon-in photon-out experimental technique that measures the scattering cross section of the material. Figure 3.5 shows a schematic picture of an x-ray scattering experiment. An incident photon with wavevector $\boldsymbol{k}_i$, polarization ϵ_i, energy E_i impinges on the sample and gets scattered. After the scattering process, the outgoing, scattered photon has the wavevector $\boldsymbol{k}_f$, polarization ϵ_f, energy E_f that can either be equal to those of the incident photon or, more generally, different. During the scattering process, the momentum and energy conservation rules hold.

$$\text{Momentum conservation}: \boldsymbol{Q} = \boldsymbol{k}_f - \boldsymbol{k}_i$$
$$\text{Energy conservation}: \hbar\omega = E_f - E_i$$

The information measured from the scattering experiment is expressed in terms of the momentum transfer $\boldsymbol{Q}$ and the energy transfer $\hbar\omega$. The momentum transfer is determined by the relative orientation between the incident beam and the scattered beam and the orientation of the sample determines the interference between incoming and outgoing beams. It is practically very demanding to change the direction of the incident beam, so it is usually fixed throughout the entire experiment. Instead, angular degrees of freedom are given to the sample and the detector that measures the scattered photon.

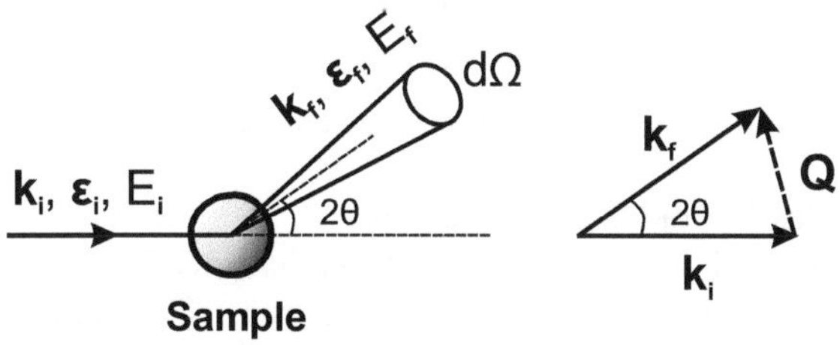

Fig. 3.5 Schematic picture of an x-ray scattering experiment. An incident photon with the wavevector $\boldsymbol{k}_i$, polarization ϵ_i, energy E_i is scattered by the sample. The scattered photon has the wavevector $\boldsymbol{k}_f$, polarization ϵ_i, energy E_f. The scattering angle is characterized by the momentum transfer $\boldsymbol{Q} = \boldsymbol{k}_f - \boldsymbol{k}_i$

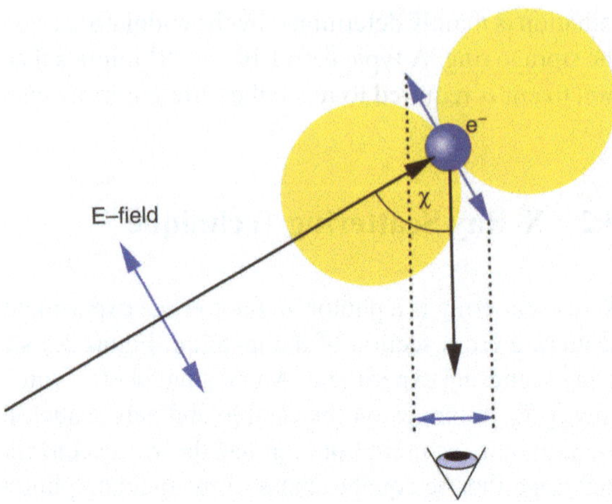

Fig. 3.6 Schematic picture of an electromagnetic radiation induced by an oscillating dipole, seen by an observer. Reproduced with permission of John Wiley & Sons—Books, from [1]

3.2.1 Thomson Scattering

The basic feature of X-ray scattering is captured by classical Thomson scattering. An oscillating electric field accelerates an electron that acts as a dipole antenna (Fig. 3.6). This oscillating dipole produces a time-dependent electric and magnetic field that radiates in space, i.e. electromagnetic radiation. A distant observer sees a projection of the time-dependent electric field generated by the oscillating dipole. In the case of a free electron, the intensity of the scattered electromagnetic radiation is given by the formula below.

$$\frac{d\sigma}{d\Omega} = |E(\chi)|^2 = \left|\left(\frac{e^2}{4\pi\epsilon_0 mc^2}\right)\cos\chi\right|^2 = r_0^2|\epsilon_i \cdot \epsilon_f|^2$$

e : electron charge, m : electron mass

c : speed of light, ϵ_0 : vacuum permitivity

$\epsilon_i(\epsilon_f)$: polarization of the incident(scattered) photon

r_0 is called the classical radius of the electron.

This simple result has several important implications. First of all, the primary scattering source for the x-rays in a solid is electrons. There are in fact two types of charged particles in a solid: atomic nuclei and electrons. Since the nuclei and electrons have the same number of total charge, the scattering intensity is proportional to $1/m^2$ and the scattering power of the electrons overwhelms that of the nuclei by $(\frac{m_e}{m_p})^2 \sim 10^6$. Therefore, the contribution of the nuclei in terms of the scattering cross section is completely ignored. In addition, since the scattering amplitude is proportional to the number of electrons in an atom Z, the scattering intensity of an atom is proportional to Z^2. This explains why x-ray scattering is less sensitive to low-Z elements, such as oxygen. Lastly, having a scattering plane perpendicular to the x-ray polarization (incident σ-polarization) ensures a maximum scattering intensity in the case of elastic scattering ($\cos\chi = 1$).

3.2.2 *Dynamical Structure Factor*

In general, the x-ray scattering is a measurement of a material-specific dynamical structure factor $S(\boldsymbol{Q}, \omega)$ multiplied by the Thomson scattering factor.

$$\frac{d^2\sigma}{d\Omega dE} = \left(\frac{k_f}{k_i}\right) r_0^2 |\epsilon_f \cdot \epsilon_i|^2 S(\boldsymbol{Q}, \omega)$$

A detailed formula of the dynamical structure factor can be given by Fermi's golden rule $\Gamma_{i\to f} = \frac{2\pi}{\hbar} \left|\langle f|H'|i\rangle\right|^2 \rho(E_f)$. In order to get the interaction Hamiltonian H', we start with the Hamiltonian of a non-relativistic electron with an electromagnetic field.

$$H = \frac{1}{2m}\left(\boldsymbol{p} - \frac{e}{c}\boldsymbol{A}\right)^2 + V(\boldsymbol{r}) \tag{3.1}$$

$$= \left[\frac{\boldsymbol{p}^2}{2m} + V(\boldsymbol{r})\right] - \frac{e}{2mc}(\boldsymbol{p}\cdot\boldsymbol{A} + \boldsymbol{A}\cdot\boldsymbol{p}) + \frac{e^2}{2mc^2}\boldsymbol{A}^2 \tag{3.2}$$

Apart from the Hamiltonian in the absence of the field $H_0 = \frac{\boldsymbol{p}^2}{2m} + V(\boldsymbol{r})$, two interaction terms arise from the light-matter interaction.

$$H_1' = -\frac{e}{2mc}(\boldsymbol{p}\cdot\boldsymbol{A} + \boldsymbol{A}\cdot\boldsymbol{p}) \tag{3.3}$$

$$H_2' = \frac{e^2}{2mc^2}\boldsymbol{A}^2 \tag{3.4}$$

3.2.2.1 Non-resonant Inelastic X-Ray Scattering Case

Non-resonant inelastic x-ray scattering arises from the first-order perturbation of the term $H_2' = \frac{e^2}{2mc^2}\boldsymbol{A}^2$. Using Fermi's golden rule, the dynamical structure factor reads,

$$S(\boldsymbol{Q}, \omega) = \sum_{i,f} |\langle \boldsymbol{k}_f, f|\boldsymbol{A}^2|\boldsymbol{k}_i, i\rangle|^2 \delta(E_f - E_i + \hbar\omega) \tag{3.5}$$

The vector potential $\boldsymbol{A} \sim \sum_{k\alpha}\left(\boldsymbol{a}_{k\alpha}^{+} e^{i\boldsymbol{k}\cdot\boldsymbol{r}} + \boldsymbol{a}_{k\alpha} e^{-i\boldsymbol{k}\cdot\boldsymbol{r}}\right)$ selects $e^{i\boldsymbol{Q}\cdot\boldsymbol{r}}$ in the interaction term and the equation is rewritten as follows.

$$S(\boldsymbol{Q},\omega)=\sum_{i,f}|\langle f|e^{i\boldsymbol{Q}\cdot\boldsymbol{r}}|i\rangle|^2\delta(E_f-E_i-\hbar\omega) \tag{3.6}$$

where $|\boldsymbol{k}_i, i\rangle$ ($|\boldsymbol{k}_f, f\rangle$) refers to the combined state of the photon with the wavevector $\boldsymbol{k}_i(\boldsymbol{k}_f)$ and the electron state $|i\rangle$ ($|f\rangle$).

Van Hove [3] showed that Eq. (3.6) leads to the Fourier transform of the density-density correlation function.

$$S(\boldsymbol{Q},\omega)=\frac{1}{2\pi}\int dte^{-i\omega t}\langle\rho(\boldsymbol{Q},0)\rho^*(\boldsymbol{Q},t)\rangle \tag{3.7}$$

Equation 3.7 provides a basic framework to calculate the non-resonant inelastic x-ray scattering spectra. Assuming a harmonic approximation of the phonons along with the detailed information of the phonon eigenvectors in the system[1] enables an analytic calculation of Eq. 3.7.

3.2.2.2 Resonant Inelastic X-Ray Scattering Case

When the incident photon energy is close to an absorption edge, the term $H_1' = -\frac{e}{2mc}(\boldsymbol{p}\cdot\boldsymbol{A}+\boldsymbol{A}\cdot\boldsymbol{p})$, whose first order perturbation gives an x-ray absorption cross section, becomes important (using the Coulomb gauge $\nabla\cdot\boldsymbol{A}=0$, $\boldsymbol{p}\cdot\boldsymbol{A}=\boldsymbol{A}\cdot\boldsymbol{p}$ and $H_1' = -\frac{e}{mc}(\boldsymbol{p}\cdot\boldsymbol{A})$). Since the x-ray scattering is a process that involves two photons, resonant inelastic x-ray scattering (RIXS) is described by the second-order perturbation of the term $H_1' = -\frac{e}{mc}(\boldsymbol{p}\cdot\boldsymbol{A})$.

$$S(\boldsymbol{Q},\omega)\sim\sum_f\left|\sum_n\frac{\langle f|\boldsymbol{p}\cdot\boldsymbol{A}|n\rangle\langle n|\boldsymbol{p}\cdot\boldsymbol{A}|i\rangle}{E_i-E_n}\right|^2$$

where $|i\rangle$ is the initial state and the sum is done over the possible intermediate ($|n\rangle$) and final states ($|f\rangle$). E_i (E_n) is the energy of the initial (intermediate) state.

In the RIXS process, the initial (final) state has the electronic state $|i\rangle$ ($|f\rangle$) and the photon with the wavevector $\boldsymbol{k}_i$ ($\boldsymbol{k}_f$), polarization ϵ_i (ϵ_f), and the energy $\hbar\omega_1$ ($\hbar\omega_2$). Then, the dynamical structure factor reads

[1] Density functional theory (DFT) provides a way to calculate the electronic band structure of solids. Once the electronic structure is available, phonon dispersion of the given crystal can be calculated using e.g. frozen phonon method or linear response method. The detailed discussion on these methods is beyond the scope of this thesis and can be found in various reviews [4–6].

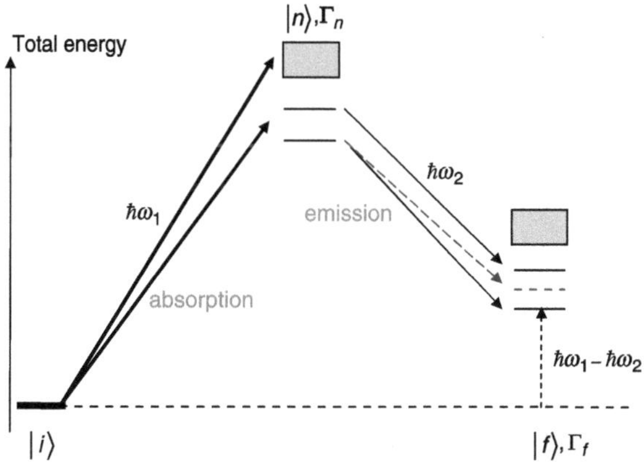

Fig. 3.7 Schematic picture of the RIXS scattering process. Reproduced with permission of Cambridge University Express through PLSclear, from [7]

$$S(\boldsymbol{Q},\omega) \sim \sum_f \sum_{j,j'} \left| \sum_n \frac{\langle f|(\epsilon_f^* \cdot \boldsymbol{p}_{j'})e^{-i\boldsymbol{k}_f \cdot \boldsymbol{r}_{j'}}|n\rangle\langle n|(\epsilon_i \cdot \boldsymbol{p}_j)e^{i\boldsymbol{k}_i \cdot \boldsymbol{r}_j}|i\rangle}{E_i + \hbar\omega_1 - E_n - i\Gamma_n} \right|^2 \times \delta(E_f - E_i - \hbar(\omega_1 - \omega_2)) \tag{3.8}$$

where the enhancement of the scattering cross section in the resonant x-ray scattering is demonstrated by the diverging denominator at the absorption edge.

Figure 3.7 shows the schematic energy diagram of the RIXS scattering process described in Eq. 3.8. An incident photon with the energy $\hbar\omega_1$ is absorbed to cause an excitation of an electron from the state $|i\rangle$ to the intermediate state $|n\rangle$. The intermediate state can be either undisturbed (direct RIXS) or disturbed (indirect RIXS) by the strong core hole potential that scatters the valence electrons within the core-hole lifetime, as illustrated in Fig. 3.8. After the core-hole lifetime, a secondary photon with the energy $\hbar\omega_2$ is emitted following the second transition of the electron from the intermediate state $|n\rangle$ to the final state $|f\rangle$. Equation 3.8 provides a physical ground to calculate RIXS spectra. The detailed discussion on calculation methods is beyond the scope of this thesis and can be found in the review [8].

3.2.3 Experimental Setup

The experimental synchrotron data presented in this thesis were measured at two beamlines of the European Synchrotron (ESRF): ID28 beamline for non-resonant inelastic x-ray scattering (Chap. 4) and ID32 beamline for resonant inelastic x-ray scattering (Chap. 5).

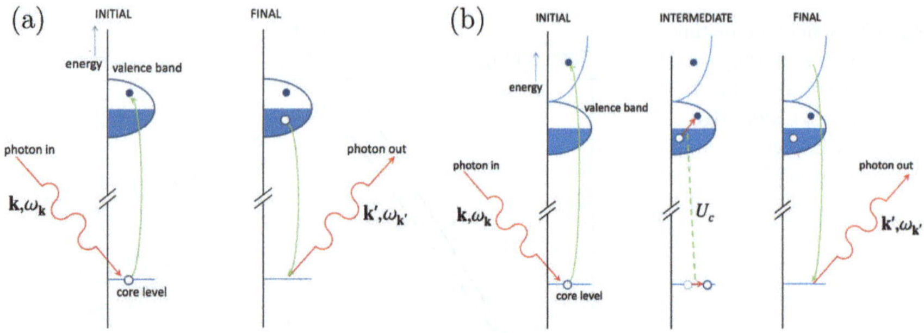

Fig. 3.8 Schematic picture of direct and indirect RIXS processes. Reprinted figure with permission from [8]. Copyright (2011) by the American Physical Society. **a** In a direct RIXS process, an incident photon with the wavevector $\boldsymbol{k}$ and the energy $\hbar\omega_{\boldsymbol{k}}$ excites an electron from the core level to the empty valence band. Then, the empty core hole is filled by an electron in the valence band, emitting a photon with the wavevector $\boldsymbol{k'}$ and the energy $\hbar\omega_{\boldsymbol{k'}}$. **b** In an indirect RIXS process, an incident photon with the wavevector $\boldsymbol{k}$, the energy $\hbar\omega_{\boldsymbol{k}}$ excites an electron to an empty valence band. Then, the strong Coulomb interaction U_c between the core hole and valence electrons creates an excitation of a valence electron. Finally, the excited electron fills the core hole and emits a photon with the wavevector $\boldsymbol{k}'$, the energy $\hbar\omega_{\boldsymbol{k}'}$

3.2.3.1 Non-resonant Inelastic X-Ray Scattering: ID28 Beamline at the ESRF

The ID28 beamline at the ESRF is one of the few beamlines in the world that can perform high-resolution (down to meV) inelastic x-ray scattering experiments. It is specialized in the detailed investigation of phonons in crystals under extreme conditions, such as cryogenic temperatures and/or hydrostatic pressure. In order to perform our experiments, we made a series of modifications to the experimental setup to implement the uniaxial strain device.

Figure 3.9 shows the optical layout of the ID28 beamline. Among the optical elements illustrated in Fig. 3.9, several features are worth noting, since they are closely tied to the nature of the IXS experiments. First of all, high-resolution IXS experiments require a high photon flux due to their photon-hungry nature. Undulator radiation provides x-rays with much higher brilliance, especially with a narrower bandwidth, compared to bending-magnet radiation and therefore is employed for IXS experiments.

Having a well-monochromatized beam is of utmost importance in the IXS experiments since it is crucial in determining the total energy resolution. In order to reach ~1 meV bandwidth, ID28 features a special monochromator in nearly full backscattering geometry for the following reason. Differentiating Bragg's law ($2d \sin\theta = \lambda$) while assuming a perfect crystal ($\Delta d/d = 0$) gives $\Delta\lambda/\lambda = \cot\theta\Delta\theta$. This implies that getting close to the back-scattering geometry ($\theta \rightarrow 90°$) minimizes the beam-divergence-originated energy bandwidth of the beam ($\Delta\lambda/\lambda$) by setting $\cot\theta$ close to zero. In this limit, it is the Darwin width of the specific Bragg reflection used for the crystal element of the monochromator that determines the energy bandwidth. The

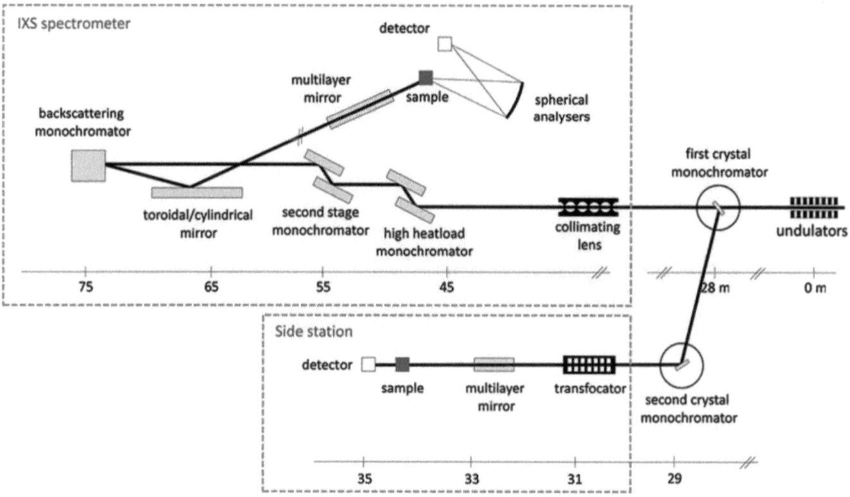

Fig. 3.9 General optical layout of the ID28 beamline in the ESRF. Reproduced with permission of the International Union of Crystallography, from [9]

ID28 beamline uses the Bragg reflections of a high-quality Silicon crystal, from (7, 7, 7) to (13, 13, 13). The higher the Bragg reflection order is, the higher the energy resolution, but there is a trade-off with a correspondingly lower flux.

This argument holds identically for the crystal analyzer mounted on the IXS spectrometer downstream from the monochromator. As a compromise between the momentum resolution and the maximum signal acquisition, the angular acceptance of the crystal analyzer in the spectrometer was set to $\sim 4 \times 10\,\text{mrads}^2$ in this beamline. This requires a larger, spherical crystal analyzer that covers this angular range, different from monochromators that accept x-rays with $10^{-2} - 10^{-3}$ mrad divergence. This beamline is equipped with 9 spherical crystal analyzers that can measure 9 different $\boldsymbol{Q}$-points simultaneously, each of which was fabricated by gluing 12000 small crystals on a spherical silicon substrate.

The energy of the scattered beam is then resolved by temperature control of the crystal analyzer that utilizes the thermal expansion of silicon. The variation of the lattice spacing at different temperatures selects a certain wavelength from the scattered beam based on Bragg's law. The high stability of the temperature control of the analyzer ($\Delta T < 0.0006\,°\text{C}$) ensures a reliable measurement of the spectral energy over the relatively long acquisition time (Fig. 3.10).

3.2.3.2 Resonant Inelastic X-Ray Scattering: ID32 Beamline at the ESRF

The ID32 beamline in the ESRF is a soft x-ray spectroscopy beamline, where one branch hosts the ERIXS spectrometer dedicated to RIXS experiments. The spectral

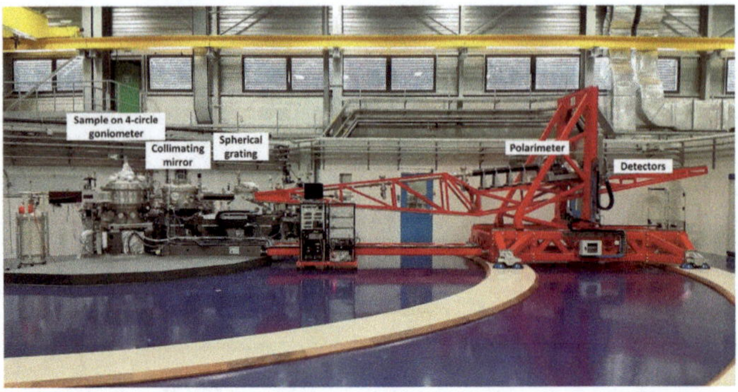

Fig. 3.10 The high-resolution ERIXS spectrometer at the ID32 beamline. Reprinted figure with permission from [10]

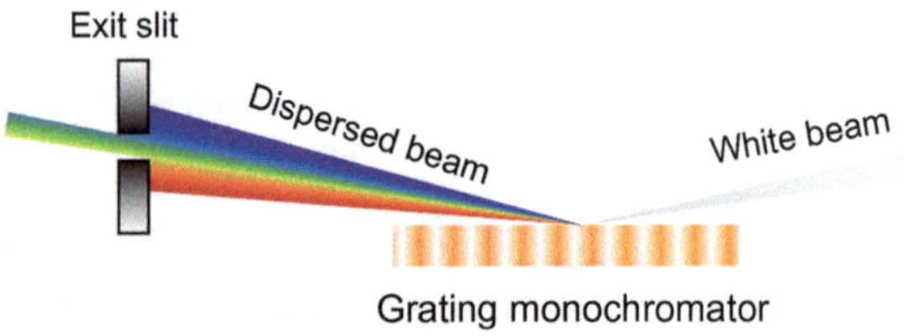

Fig. 3.11 Schematic picture that shows how the grating monochromator combined with an exit slit produces x-rays with a narrow bandwidth

range of the incident photons provided by the beamline is 0.3–1.6 keV, which includes the $L_{2,3}$-edges of the 3*d* transition metals and the $M_{4,5}$-edges of the 4*f* rare-earth elements. The medium energy resolution in routine operations reaches ∼60 meV at Cu-L_3 edges, but the combined energy resolution of the setup can be pushed down to <35 meV resolution at the expense of the incident flux. This is by far the best energy resolution obtained in the soft x-ray energy ranges to date.[2]

As in the case of IXS experiments, RIXS experiments are very photon-hungry and therefore need high flux from an undulator. Unlike the crystal monochromator used in the IXS beamline, soft x-ray beamlines utilize a plane grating monochromator (PGM) with variable line spacing (VLS). The PGM disperses the white beam coming from the undulator in different angles and a selected spectral range of x-rays passes through an exit slit, as shown in Fig. 3.11. With a fixed size of the exit slit, the bandwidth of the incident beam is a monotonous function of the distance between the monochromator and the exit slit, which is why the ID32 beamline has such a long optical path.

Spectroscopic experiments with soft x-rays are usually carried out under ultra-high vacuum due to the strong attenuation in air. This puts constraints on the sample environment, e.g. restricting the angular degrees of freedom. The ERIXS spectrom-

---

[2] At the time of writing, the Centurion RIXS spectrometer of the SIX beamline at NLSLII Brookhaven claims a new record resolution of 25 meV, but this was reached until now only in commissioning.

eter at the ID32 beamline is equipped with an in-vacuum four-circle diffractometer manipulator that allows one to explore most of the first Brillouin zone in high-T_c cuprates.

ERIXS has an 11 m-long arm that moves in the scattering plane covering $2\theta = 50-150°$, as shown in Fig. 3.10, and it is connected to the sample chamber through a differentially pumped ribbon section, which keeps it in vacuum even while moving the 2θ arm. Incident photons scattered from the sample are collimated by an elliptical horizontal mirror and dispersed in energy by a spherical VLS grating analyzer. The scattered beam travels the 11 m distance and is measured by the CCD detector. Photons with different energies are seen as isoenergetic lines at different heights in the CCD detector.

3.3 Uniaxial Strain Technique

3.3.1 Piezoelectric-Based Strain Device

Experiments presented in this thesis were performed with the modified versions of the piezoelectric-based strain device already used in the studies of superconducting Sr_2RuO_4 by Hicks *et al.* [12]. Figure 3.12 shows a schematic picture of the strain device used in our non-resonant inelastic x-ray scattering experiments. An opening of 90° on the back side was intentionally fabricated for the transmission of the scattered photons.

A detailed picture of the strain device with the mounted sample is displayed in Fig. 3.13. Uniaxial stress is produced by the piezoelectric stacks present on the sides of the strain device. Selective application of voltage on individual piezoelectric stacks enables one to apply tensile and compressive stress to the sample. In order to apply compressive stress, positive voltages are applied on the two piezoelectric stacks at the center (dashed arrows in the Fig. 3.13). The two stacks are then elongated and push the central part of the device inward, resulting in uniaxial pressure on the sample at

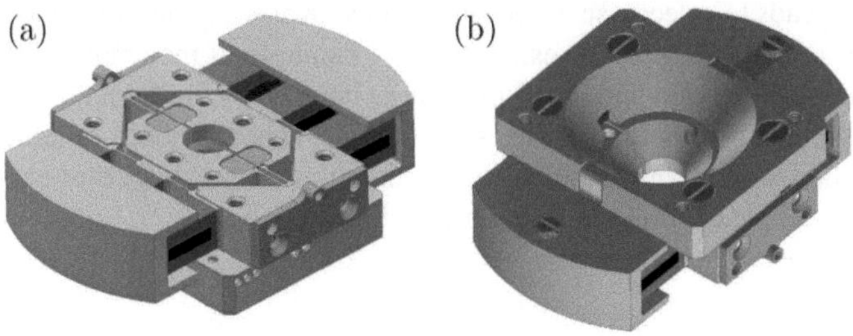

Fig. 3.12 Strain device used for non-resonant inelastic x-ray scattering experiments. **a** Front side **b** Back side with an opening for transmission of x-ray. From [11]. Reprinted with permission from AAAS

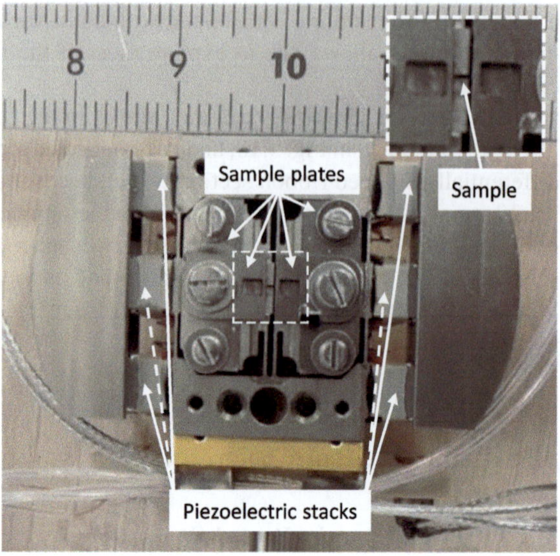

Fig. 3.13 Photo of the strain device taken before the x-ray scattering experiment. The black needle at the center of the device is the sample mounted with the help of the sample plates. Six piezoelectric stacks are present on the sides. The inset with the white dashed frame shows a zoom of the central part, where the sample is mounted

the center. On the other hand, when the voltages are applied on the four piezoelectric stacks on the outer sides (solid arrows in Fig. 3.13), this pulls the central part and leads to tensile strain. Voltage application on the strain device merely requires an electrical connection, which can be easily implemented in a cryostat. This allows *in-situ* control of the sample strain at cryogenic temperatures.

The strain device is equipped with two parallel-plate-capacitor-based displacement sensors connected in parallel. Each side of the parallel plate capacitor is attached to the different blocks inside the strain device that are decoupled from each other. Typical dimensions of the parallel plate capacitor are A(area) $= 1.6\times 2.8\,\text{mm}^2$ and d(distance) $= 50\,\mu$m. This gives a capacitance value of $2\times 8.854\times A[\text{mm}^2]/d[\mu\text{m}]$ $= 1.6$ pF. The capacitance bridge AH2550A from Andeen-Hagerling that we used has a precision of $\sim 10^{-6}$ pF, which can resolve a displacement smaller than 1Å.

When compressive stress is applied, the displacement of the capacitance sensor decreases, resulting in an increase of the capacitance ($C = \frac{\epsilon_0 A}{d}$). Tensile strain, on the other hand, leads to a decrease of the capacitance. Knowing the sample length under strain, the sample strain can be estimated. For example, if the length of the sample under strain is $L = 1$ mm and the displacement induced by the device is $\Delta L = 1\,\mu$m, then the estimated sample strain is $\frac{\Delta L}{L} = 0.1\%$. This provide a reliable estimate of the applied strain, which will be independently verified via other experimental techniques, as discussed in the following section.

3.3.2 Characterization

This section presents selected examples of several experiments that I did to characterize the stress application by the piezoelectric-based strain device. The experiments were carried out with two experimental instruments available in the labs at the Max Planck Institute for Solid State Research: X-ray diffraction and Raman spectroscopy. X-ray diffraction experiments were also performed at the high-pressure diffraction beamline ID27 of the ESRF. Modifications of the experimental setup to implement the strain device, such as design of the device holder and installation of the electronic connections inside the cryostat, were done with the help of technical staff.

3.3.2.1 Diffraction

The best way to confirm the sample strain is to check the variation of the lattice constants by means of x-ray diffraction. The x-ray tube with Molybdenum target at Max Planck Institute Stuttgart produces a collimated beam with a cross-section of $0.8 \times 0.7\,\text{mm}^2$ (FWHM × FWHM) that contains the characteristic x-ray lines $K_{\alpha1}$ ($\lambda = 0.7093$ Å, E = 17.48 keV) and $K_{\alpha2}$ ($\lambda = 0.7136$ Å, E = 17.37 keV). The K_{β} ($\lambda = 0.632$ Å, E = 19.617 keV) components are removed from the outgoing x-ray beam by the setup optics. $K_{\alpha1}$ and $K_{\alpha2}$ are too close to each other in energy and cannot be selected by the optics.

In this in-house setup, the main limitation of the angular resolution, which directly limits also the strain resolving capability, is the x-ray beam size on the sample. The horizontal FWHM of the x-ray beam 0.8 mm gives an angular acceptance of 0.8/800 = 1 mrad = 0.057°, when the detector is located ~80 cm away from the sample. An extremely sharp Bragg reflection (e.g. δ-function) will then be seen as a peak with FWHM = 0.057° in the diffraction pattern. When measuring $YBa_2Cu_3O_{6.67}$, a b-axis compression of 0.1% causes a 2θ shift of the (0 2 0) Bragg reflection by 0.02° ($2\theta \sim 21°$). This is almost half of the angular acceptance and therefore, our instrument is expected to observe strain effects larger than or equal to 0.1%.

Figure 3.14a shows typical θ–2θ scans of the (0 2 0) Bragg reflection of $YBa_2Cu_3O_{6.67}$ at different strain levels. Each scan has two peaks owing to the incident beam having two wavelengths from $K_{\alpha1}$ and $K_{\alpha2}$. Since they are well-separated in our diffraction scan, it does not cause any difficulty in distinguishing the shift of the Bragg peak.

Our measurement shows a clear, systematic shift of the Bragg peak position under uniaxial strain. Compressive strain leads to a decrease of the lattice constant and Bragg's law ($2d \sin\theta = \lambda$, $\sin\theta \propto 1/d$) allows us to estimate the lattice constant variation from the Bragg peak position. Figure 3.14b compares the nominal strain, which is obtained from the displacement of the piezoelectric stack measured by the internal capacitance sensor, and the actual strain measured via diffraction. The two strain scales agree reasonably well with each other, confirming the strain effect by the piezoelectric-based strain device.

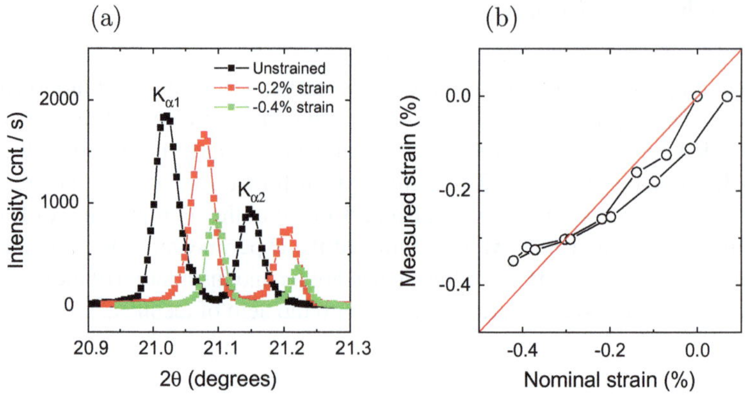

Fig. 3.14 **a** Scans of the (0 2 0) Bragg reflection at different compressive strains **b** Comparison between nominal strain from the capacitance measurement and actual strain from the Bragg reflection shift. The red line corresponds to the case when nominal strain equals measured strain

Similar experiments were performed using synchrotron radiation. The much higher brilliance (i.e. smaller beam size and angular spread) of the synchrotron radiation enabled us to observe how strain is distributed across the sample. Figure 3.15a shows the strain value measured at different positions of the sample and compares the result measured without strain and at -0.12% strain. The strain level of -0.12% does not show any inhomogeneous distribution across the sample. The simultaneous measurement of both b- and c-lattice constants under strain provides us a Poisson's ratio $\nu_{yz} = 0.25 \pm 0.02$, as shown in Fig. 3.15b. Note that the variation of the lattice constant is quite linear with the nominal strain. This implies that neither the strain device nor the sample and the epoxy deviated from the elastic limit during the strain application.

3.3.2.2 Raman Scattering

Another way to check the strain effect is to monitor the excitations that arise from the lattice, i.e. phonons. External pressure results in a decrease of the lattice constant, which leads to an increase of the effective spring constant between atoms and thus an increase of the phonon energy ($\omega = \sqrt{k/m}$).

Raman scattering is one of the most widely used spectroscopic tools to study zone-center phonons in a solid. A variation of the lattice parameter is reflected in a shift of the phonon energy in the Raman spectrum. Figure 3.16a shows how the energy of different phonons in YBCO changes as a function of hydrostatic pressure. For all the investigated phonons, the phonon energy increases linearly as the pressure increases. The Grüneisen parameters $\gamma = \frac{d\omega/\omega}{dV/V}$, which yield the ratio between the phonon energy change and the volume change, are shown in Fig. 3.16b.

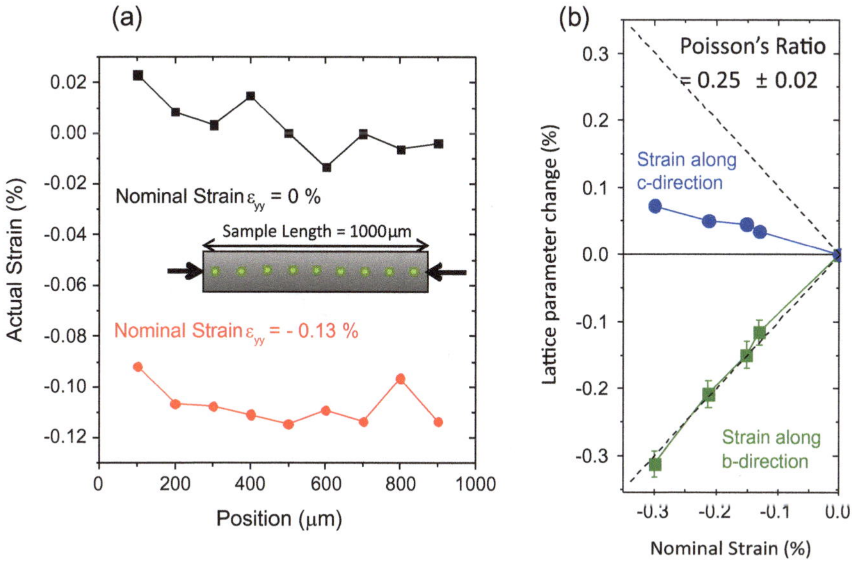

Fig. 3.15 **a** Comparison of the strain distribution between unstrained condition and −0.13% strain. *b*-axis compression was applied on the $YBa_2Cu_3O_{6.67}$ and strain was measured at different positions on the sample with steps of 100 microns. **b** Lattice parameter change of *b* and *c* under *b*-axis compression

a

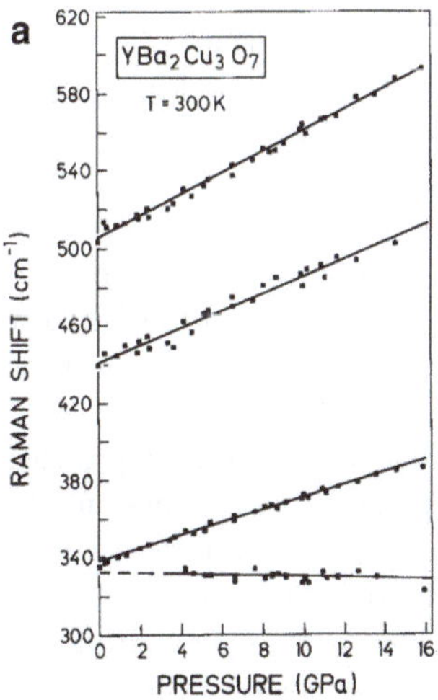

b

TABLE 1
Pressure coefficients and Grüneisen parameters for Raman modes in $MBa_2Cu_3O_7$ (M = Y, Eu)

ω_0 (cm^{-1})	dω/dP (cm^{-1}/GPa)	γ(exp)	γ(calc)
$YBa_2Cu_3O_7$			
145	1.3 ± 0.2	1.5	1.6
338	3.2 ± 0.2	1.6	1.6
440	4.4 ± 0.3	1.7	1.7
504	5.5 ± 0.2	1.9	1.3
~330	-0.5 ± 0.3	- 0.3	
$EuBa_2Cu_3O_7$			
313	2.5 ± 0.3	1.4	1.6
436	3.2 ± 0.3	1.3	1.7
506	6.5 ± 0.3	2.2	1.3

Fig. 3.16 **a** Pressure dependence of the energy of different phonon modes measured in $YBa_2Cu_3O_7$. **b** Table of the pressure coefficients and Grüneisen parameters measured for different phonon modes. Reprinted figure from [13]

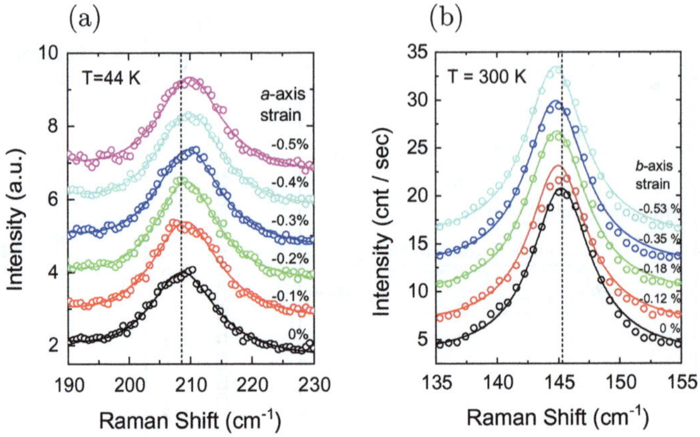

Fig. 3.17 **a** Low-temperature phonon at $\sim$210 cm^{-1} measured by Raman in $YBa_2Cu_3O_{6.67}$ under uniaxial pressure along the a-axis. The spectra were measured at T = 44 K. **b** Raman spectra of the out-of-plane vibration mode of the Cu in the CuO_2 plane in $YBa_2Cu_3O_{6.67}$ under unaxial pressure along the b-axis. The spectra were measured at room temperature

Note that the Grüneisen parameters are of the order of 1, which means $\Delta\omega/\omega \sim \Delta V/V$. This can be used to estimate the effect of uniaxial pressure on the phonon energy. Under uniaxial strain of 1%, the phonon energy shift is expected to be around 1%. Considering a typical phonon energy of a few hundreds of cm^{-1}, uniaxial pressure is expected to cause an energy change of the phonons by several cm^{-1} ($\lesssim$1 meV). Raman spectroscopy is capable of detecting these tiny shifts and the Raman spectrometer at Max Planck Institute Stuttgart has an energy resolution of 0.8 cm^{-1} when used with a grating with line density 1800 lines/mm.

Figure 3.17 shows the evolution of selected phonons of $YBa_2Cu_3O_{6.67}$ as a function of uniaxial pressure. One expects that uniaxial pressure leads to an increase of the phonon energy, as hydrostatic pressure does. This is shown in Fig. 3.17a. The low-temperature phonon at 210 cm^{-1} hardens under uniaxial pressure, which is consistent with the prediction.

Interestingly, some phonons are observed to soften under uniaxial pressure. Figure 3.17b shows an example of phonon softening upon b-axis compression. This phonon mode corresponds to the out-of-plane vibration of the Cu atoms in the CuO_2 planes with A_{1g} symmetry. The decrease of the phonon energy can be, to a first approximation, explained by the expansion of the out-of-plane lattice constant as a result of the in-plane uniaxial pressure, as the positive Poisson ratio ($\nu_{yz} = -d\epsilon_z/d\epsilon_y$) indicates. This leads to a decrease of the effective spring constant along the out-of-plane direction, which results in the softening of the phonon.

Another example of the strain effect is found in Fig. 3.18. Figure 3.18a shows Raman spectra of the B_{2g} phonon mode at 393 cm^{-1} measured in the iridate compound Sr_2IrO_4. A systematic hardening of the phonon mode is clearly observed, as the compressive strain increases along the [1 0 0] direction. The phonon energy

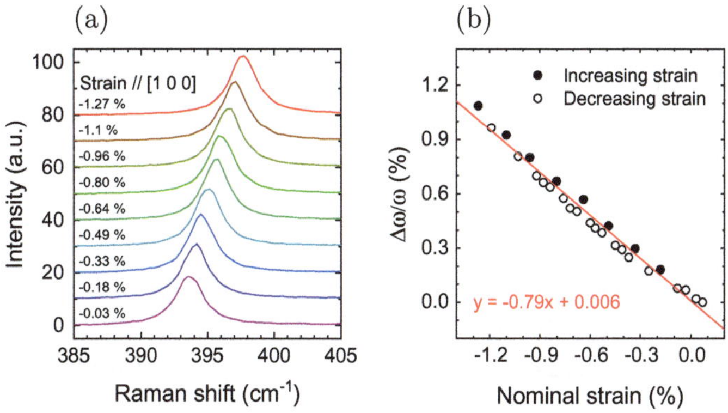

Fig. 3.18 **a** Raman spectra of the B_{2g}-phonon in Sr_2IrO_4 at 393 cm^{-1} at different uniaxial strains along the [1 0 0] direction. Spectra are shifted by vertical offsets. **b** Phonon energy change $\Delta\omega/\omega$ versus uniaxial strain along the [1 0 0] direction. The red line shows the linear fit to all the data points

change $\Delta\omega/\omega$ as a function of strain is displayed in Fig. 3.18b. The change of the phonon energy remains linear up to the nominal strain of -1% and the hysteresis observed in the cycle of increasing and decreasing the strain is reasonably small ($<0.03\%$ in strain). This provides important experimental evidence that cuprate and iridate crystals maintain a predominantly elastic behavior up to $\geq -1\%$ strain. It also implies that the nominal strain scale based on the displacement sensor is a reliable indicator of the actual strain in the sample.

3.4 Crystal Preparation

YBCO crystals were grown using the flux method that is described by Lin *et al.* [14]. This produces single crystals of YBCO with a size of >2 mm $\times$ >2 mm $\times$ <0.7 mm. As-grown YBCO crystals are usually inhomogeneous in their oxygen content, revealing multiple, broad transitions of superconductivity as shown in Fig. 3.19a.

3.4.1 Annealing

Since cuprates reveal a variety of phases depending on the hole doping of the CuO_2 planes, it is crucial to reliably control their doping level. In the underdoped case, a homogeneous doping level is characterized by a sharp superconducting transition as that shown in Fig. 3.19b. In YBCO, this is mainly obtained by annealing the sample under specific conditions with a controlled oxygen partial pressure. Lindemer

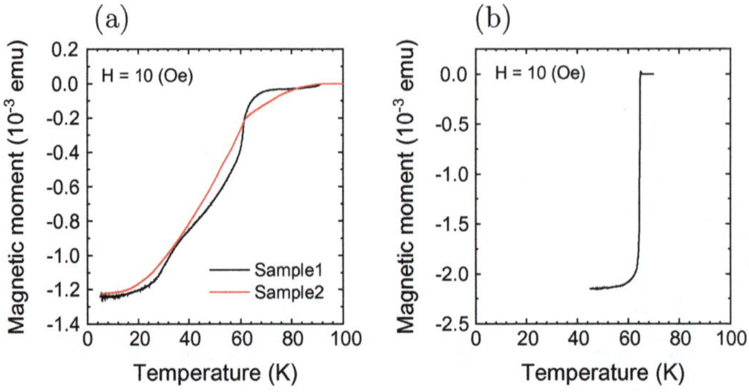

Fig. 3.19 Magnetization versus Temperature measured for **a** two as-grown samples **b** a $YBa_2Cu_3O_{6.67}$ sample. Samples were first cooled without a magnetic field and the measurement was done with field H = 10 Oe while warming up

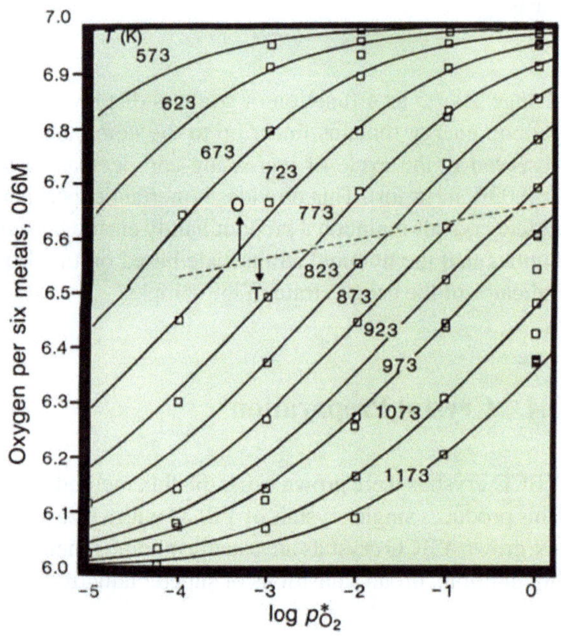

Fig. 3.20 Relation between oxygen content in YBCO, annealing temperature and oxygen partial pressure. Reprinted figure from [15]

et al. [15] characterized the oxygen stoichiometry of YBCO for different annealing conditions of temperature and oxygen partial pressure, as shown in Fig. 3.20.

There is more than one combination of parameters that results in the same level of oxygenation in YBCO. Practically, several aspects have to be considered when choosing reasonable control parameters. It is recommended to work with an annealing temperature high enough to finish the annealing procedure within a reasonable period of time. The oxygen diffusion coefficient in YBCO is well described by the Arrhenius equation $D = D_0 \exp(-\Delta E/kT)$ with $\Delta E = 1\,\mathrm{eV}$ [16], which implies that the higher the temperature, the higher the diffusion rate. However, the temperature should not be too high for the furnace in light of the risk of instabilities of the target temperature. For the preparation of $YBa_2Cu_3O_{6.67}$, the samples were annealed at 555 °C under oxygen partial pressure of 0.05 bar.

3.4.2 Detwinning

After the annealing procedure, the resulting YBCO crystals have twin domains due to orthorhombicity induced by the oxygen atoms in the chain layer. Making the crystal twin-free is important in investigating the intrinsic anisotropy of the system. An experimental method to detwin YBCO crystals is to apply uniaxial pressure at an elevated temperature. This facilitates oxygen diffusion inside the crystal and rearranges oxygens in the new potential of the chain layer due to the external compression. A typical method that I used is to apply 50–60 MPa of uniaxial pressure at a temperature of 400 °C for 2 h.

The presence of twin domains can be easily observed by means of polarized optical microscopy [17], as shown in Fig. 3.21. Depending on the relative orientation of oxygen chains with respect to the incoming light polarization, different twin domains appear as different colors. A signature of a detwinned YBCO crystal is a uniform color on the surface observed under the polarized microscope that changes under a 90° rotation. One should note that this method only provides the twinning information on the surface, due to the limited penetration depth of visible light. Nevertheless, this simple method has always been successful in producing high-quality untwinned YBCO crystals, which can be further checked in the synchrotron x-ray scattering experiments. For example, our resonant x-ray scattering experiments consistently observed a peak of the oxygen order only along H, but not along K.

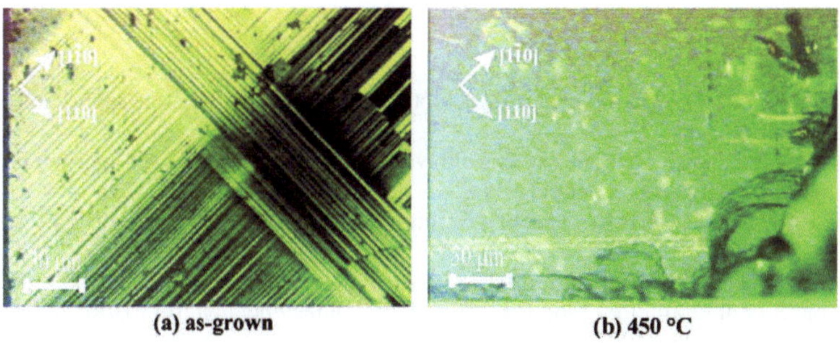

Fig. 3.21 Image of the YBCO single crystal taken under the polarized microscope (left) before detwinning (right) after detwinning. Reprinted figure from [17]

3.4.3 Preparation for Strain Experiment

Strain experiments require an intense and time-consuming procedure of crystal preparation that involves cutting and polishing the surfaces of the crystals. The crystals for strain experiments require a needle shape that is long in the direction of applied stress and short in the other directions. Practically, we aim to fabricate needle-shaped samples with a length of >1.2 mm, a width of 0.2–0.3 mm and a thickness of 0.07–0.1 mm. Our YBCO samples usually have dimensions of $\sim 2\times 2\times 0.2$–$0.5\,\text{mm}^3$. Hereafter, the detailed procedure of the sample preparation is reported:

1. Cut a large sample into 'bars' using a wire saw, so that they have the target width of 0.2–0.3 mm. The wire saw that we use (WS-22 from IBS Fertigungs- und Vertriebs-GmbH) ensures that the sample loss due to the cutting process is $\leq 100\,\mu$m.
2. Polish the bar-shaped samples down to 'needles' with the desired thickness of 0.07–0.1 mm. We polish both surfaces of the sample, in order to remove any volume near the sample surface that might be susceptible to cleaving under shear stress during uniaxial stress application.

This procedure turns out to naturally select the sample portions of greater quality out of the good YBCO crystals, as otherwise the sample typically breaks at some point during the procedure. For example, it was observed that samples with bad rocking scans in XRD (e.g. having a broad peak or two peaks in ω-scans), which indicates a large mosaicity of the crystal, tend to crumble during the polishing process and did not survive in the end. To mention another example, some samples show broad Bragg reflections of (0 0 L) in the $\theta-2\theta$ scan, indicating a weaker correlation along the c-axis than that of samples with good quality. These samples tend to easily cleave during the polishing process.

References

1. Willmott P (2011) An introduction to synchrotron radiation, techniques and applications. Wiley, New York
2. Attwood D (2016) X-rays and extreme ultraviolet radiation, principles and applications. Cambridge University Press, Cambridge
3. Van Hove L (1954) Phys Rev 95:249–262. https://doi.org/10.1103/PhysRev.95.249. Accessed 1 July 1954
4. Weyrich KH (1990) Ferroelectrics 104:183–194. https://doi.org/10.1080/00150199008223822
5. Baroni S et al (2001) Rev Mod Phys 73:515–562. https://doi.org/10.1103/RevModPhys.73.515. Accessed 2 July 2001
6. Gonze X, Lee C (1997) Phys Rev B 55:10355–10368. https://doi.org/10.1103/PhysRevB.55.10355. Accessed 16 April 1997
7. Rueff J-P (2010) Chapter 9: An introduction to inelastic X-ray scattering. Magnetism and synchrotron radiation. Springer, Berlin
8. Ament LJP et al (2011) Rev Mod Phys 83:705–767. https://doi.org/10.1103/RevModPhys.83.705. Accessed 2 June 2011

9. Girard A et al (2019) J Synchrotron Radiat 26:272–279. https://doi.org/10.1107/S1600577518016132
10. Brookes N et al (2018) Nucl Instrum Methods Phys Res Sect A: Accel Spectrom Detect Assoc Equip 903:175–192. https://doi.org/10.1016/j.nima.2018.07.001
11. Kim H-H et al (2018) Science 362:1040–1044. https://doi.org/10.1126/science.aat4708. https://science.sciencemag.org/content/362/6418/1040.full.pdf
12. Hicks CW et al (2014) Science 344:283–285. https://doi.org/10.1126/science.1248292. https://science.sciencemag.org/content/344/6181/283.full.pdf
13. Syassen K et al (1988) Phys C 153, 1 P4100 Times Cited:44 Cited References Count:5:264–265
14. Lin C et al (1992) Phys C: Supercond 195:291–300. https://doi.org/10.1016/0921-4534(92)90353-E
15. Lindemer TB et al (1989) J Amer Ceram Soc 72:1775–1788. https://doi.org/10.1111/j.1151-2916.1989.tb05978.x. https://ceramics.onlinelibrary.wiley.com/doi/pdf/10.1111/j.1151-2916.1989.tb05978.x
16. Erb A (1999) Habilitation, Université
17. Lin C, Kulakov A (2004) Phys C: Supercond 408–410. In: Proceedings of the international conference on materials and mechanisms of superconductivity. High temperature superconductors VII - M2SRIO, pp 27–29. https://doi.org/10.1016/j.physc.2004.02.022

Chapter 4
CDW in $YBa_2Cu_3O_{6.67}$ Under Uniaxial Pressure: Non-resonant Inelastic X-Ray Scattering

4.1 Background

4.1.1 Electron-Phonon Interaction and CDW in Cuprates

An increasing number of studies is pointing out the presence of strong electron-phonon coupling associated with CDW formation in cuprates. Inelastic x-ray scattering studies [1, 2] revealed that the dispersion of low-energy phonons in moderately doped YBCO shows a significant broadening at the CDW wavevector, with pronounced anomalies at the superconducting transition temperature T_c. In particular, Le Tacon *et al.* [1] observed a giant softening of the low-energy phonons at the CDW wavevector below the superconducting transition temperature. This calls for further research to deepen the understanding of the role of phonons in the CDW phenomenology.

4.1.2 Lattice Pressure Studies of YBCO

Lattice pressure decreases the distance between the atoms in a solid. In the case of YBCO, several hydrostatic pressure studies give insights into the effects of externally applied lattice pressure. High pressure neutron powder diffraction [3] showed an increase of the Cu(2) valence, i.e. an increase of the effective hole doping per Cu atom in the CuO_2 plane, under hydrostatic pressure, resulting from the reduced distance between the CuO_2 plane and the CuO chain layer. Optical studies under hydrostatic pressure [4] revealed an increase of the exchange interaction J, which might be a reflection of several effects, such as an increase of the hopping integral t, an increase of effective hole doping that reduces the exchange interaction or even possible phonon-magnon coupling. The high pressure effect on superconducting T_c

H. Kim, *Uniaxial Pressure Study of Charge Density Waves in a High-T_c Cuprate Superconductor*, Springer Theses,
https://doi.org/10.1007/978-3-030-99898-1_4

was also studied [5] and utilized to get access to the far-overdoped region of YBCO, which has not been explored previously with chemical substitution.

At the time of writing, experimental data that show how the CDW responds to hydrostatic pressure, and in general against lattice pressure, are minimal. This seems to be mainly due to the technical difficulties in hydrostatic pressure experiments. A conventional way of reaching high pressure is using a diamond anvil cell, which is composed of a gasket and two culets of diamonds on each side that are a few hundred microns in diameter. This already complicates the sample preparation. Especially in photon scattering experiments, it is difficult to control stray signals from the diamonds and/or other materials, such as a fluorescence from defects inside the diamond or Compton scattering from the diamond itself.

Only quite recently, a few pioneering experiments were carried out. Inelastic x-ray scattering experiments [6] showed that the phonon anomaly at the CDW wave vector disappears at the hydrostatic pressure of 1.5 GPa, which was interpreted as a suppression of the CDW by hydrostatic pressure. This is supported by an elastic x-ray scattering experiment, which did not detect a CDW signature at $P = 1$ GPa [7]. On the other hand, Hall measurements show conflicting results, suggesting either little change [8] or suppression [9] of the CDW under hydrostatic pressure. NMR results [10] show a weak dependence of CDW on hydrostatic pressure. Further investigations are needed to reach a consensus on the CDW response under hydrostatic pressure.

4.1.3 Uniaxial Pressure as a New Perturbation Method

In the doping range of interest (i.e., the underdoped region around $p = 0.12$), hydrostatic pressure leads to an increase of the superconducting T_c at low pressure ($P \sim 1$ GPa) [5]. On the other hand, uniaxial pressure was shown to cause different and even opposite effects to T_c depending on the pressure direction [11]. In fact, the isotropic nature of hydrostatic pressure inevitably obscures the contributions of individual effects, reflecting a net sum of opposite effects.

In this perspective, uniaxial pressure is better suited as a parameter to tune the physical properties in YBCO. Recently, Hicks *et al.* [12] observed an anisotropic response of T_c in Sr_2RuO_4 under uniaxial pressure, using the piezoelectric strain device that they developed. The small size of the device, the open space around the sample and its compatibility with cryogenic environments are well-suited for scattering experiments as well. The strain level that the device can reach goes up to 1–2%. Converted to pressure using the Young's modulus of YBCO (154.3 GPa) [13], this is approximately 1.5–3 GPa. Considering that the high pressure inelastic x-ray scattering study [6] showed a drastic change of the CDW under much lower pressure, uniaxial stress of a similar scale is expected to cause a noticeable change in the CDW.

4.2 Experiment

Inelastic x-ray scattering experiments were carried out at the beamline ID-28 of the ESRF. The spectrometer of this beamline has 9 crystal analyzers that enable acquisition of energy-resolved scattering intensity at 9 different scattering vectors simultaneously. Our experiments utilized a transmission geometry, where we measure scattered photons that passed through the sample. In this geometry, control of the sample thickness is critical for a successful measurement within a given experimental time window. Both too thin or too thick samples are not adequate, since they require a long averaging time to obtain statistically reasonable photon counts. Indeed, too thin samples provide only a small probing volume, while too thick samples absorb most of the incident/scattered photons.

In order to maximize the scattering intensity, a conventional approach is to prepare a sample with a thickness comparable to its absorption length. The absorption length of photons with an energy of 17.794 keV in $YBa_2Cu_3O_{6.67}$ is estimated to be around $10\,\mu$.[1] For the experiment, the sample was first polished down to $120\,\mu$m and then further thinned down using a focused ion beam. The resulting sample is depicted in Fig. 4.1.

We chose the a-axis as direction of the uniaxial pressure for various reasons. In underdoped $YBa_2Cu_3O_{6+x}$, various oxygen orders in the chain layer are found depending on the oxygen doping level. These orders have distinct patterns and periodicities along the a-axis and they translate into intense Bragg reflections along H in the reciprocal space.

In particular, $YBa_2Cu_3O_{6.67}$ displays an ortho-VIII order in the chain layer, which gives strong elastic peaks at $H = 3/8, 5/8$. This specifically disturbs the measurement of the CDW along the H-axis, where the CDW peak is observed at $H \sim 0.31$, quite close to $H = 3/8 = 0.375$. The scattering intensity from the CDW is very small and it can be easily shadowed by any nearby elastic signals in the reciprocal space. The presence of the ortho-VIII order in the (H 0 L) scattering plane, which obscures

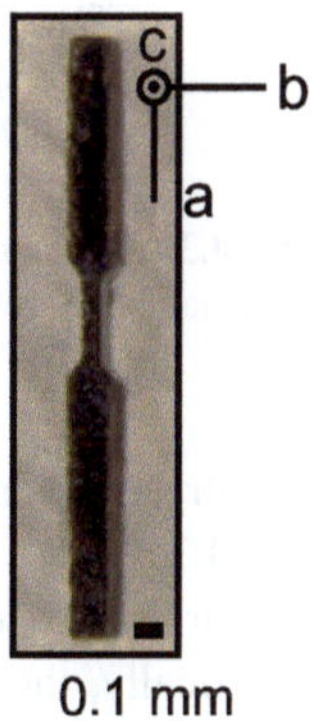

Fig. 4.1 A needle-shaped sample prepared for strain experiments. The central part of the sample was further thinned down using the focused ion beam technique. From [14]. Reprinted with permission from AAAS

[1] This can be easily calculated at the website http://henke.lbl.gov/optical_constants/atten2.html.

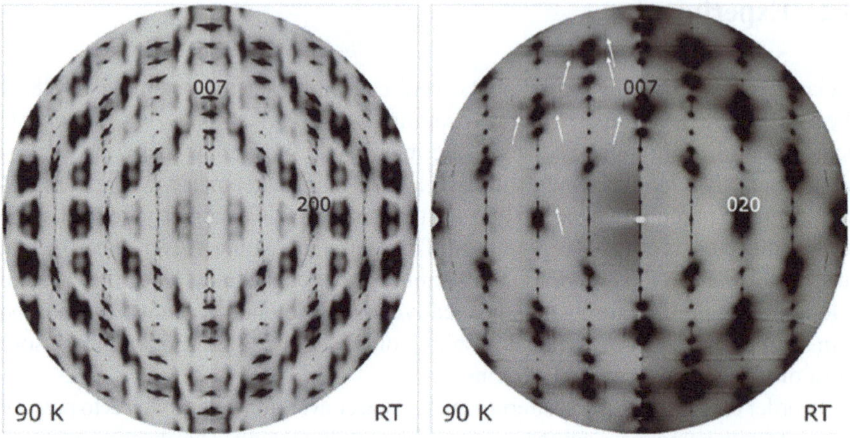

Fig. 4.2 (Left) Reciprocal space map of (H 0 L) plane in $YBa_2Cu_3O_{6.6}$ (Right) Reciprocal space map of (0 K L) plane in $YBa_2Cu_3O_{6.6}$. For each figure, the left half of the image was taken at T = 90 K, while the right half was taken at room temperature. Reprinted by permission from Nature Physics [1]

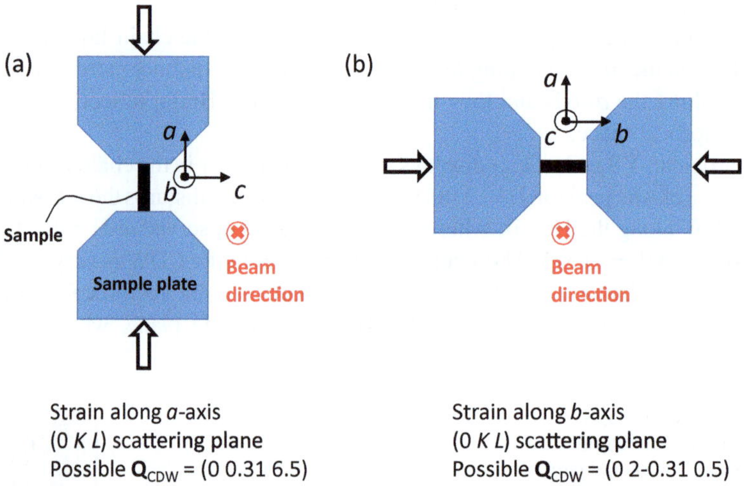

Fig. 4.3 **a** Schematic picture of the scattering geometry when strain is applied along *a*-axis **b** Schematic picture of the scattering geometry when strain is applied along *b*-axis

the already-weak CDW signal, was illustrated by Le Tacon *et al.* [1], as shown in Fig. 4.2.

For this reason, we chose to investigate the (0 K L) plane of the reciprocal space. Practically, there are two ways of establishing (0 K L) as the scattering plane, as shown in Fig. 4.3. The first is to make a long *a*-axis needle and put it perpendicular to the scattering plane (Fig. 4.3a). This approach enables us to apply a stress along the

a-axis and study the CDW and phonons at $\boldsymbol{Q}$ vectors with high L, e.g. around $\boldsymbol{Q}_{\mathrm{CDW}}$ = (0 0.31 6.5). Another way is to make a long b-axis needle and mount it parallel to the scattering plane. With this method, one can apply uniaxial pressure along the b-axis and get access to inelastic signals at $\boldsymbol{Q}$ with high K, e.g. around $\boldsymbol{Q}_{\mathrm{CDW}}$ = (0 2–0.31 0.5).

Both approaches are experimentally feasible, but we chose the geometry of a-axis strain (Fig. 4.3a) over that of b-axis strain (Fig. 4.3b). In the geometry for b-axis strain, the sample plates lie within the (0 K L) scattering plane with the risk to possibly block either incoming or scattered photons at some intermediate angles of rotation. The a-axis strain geometry is free from this risk and only restricted by the designed angular opening of the device = 45°. This is well below our 2θ of interest $\sim$20°.

In addition, the previous IXS study of Le Tacon *et al.* [1] was carried out using this geometry. This provides us a guide and a reference for the features observed in the inelastic spectra of underdoped YBCO in the absence of strain.

Last but not least, underdoped $YBa_2Cu_3O_{6.67}$ has an orthorhombic crystal structure with $a \sim 3.82$ Å $< b \sim 3.88$ Å. Applying uniaxial pressure along the a-axis makes the crystal structure more orthorhombic. This leads to a more extreme condition for the underlying electronic system and is expected to grant access to a new region in the phase space of cuprates, which is impossible to study in the bulk crystal under ambient pressure.

4.2.1 *Typical IXS Spectrum*

Figure 4.4 shows a typical spectrum of $YBa_2Cu_3O_{6.67}$ measured in our experiment. This spectrum was measured at the wavevector $\boldsymbol{Q}$ = (0 0.27 6.5). In the energy range between 0 and 25 meV, several distinct peaks are observed. They can be assigned to elastically scattered intensity and phonons, both of which are convoluted with the energy resolution function.

A scan of the elastic intensity vs momentum around $\boldsymbol{Q}_{\mathrm{CDW}}$ provides the correlation length (inverse of the peak half-width) of the CDW domains along the scan direction.

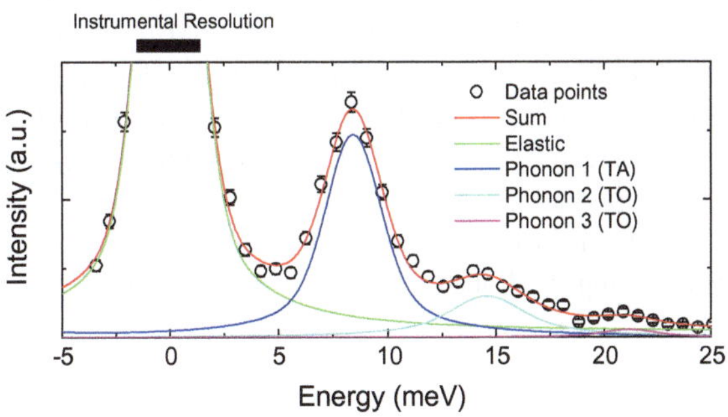

Fig. 4.4 Inelastic spectrum of $YBa_2Cu_3O_{6.67}$ at $\boldsymbol{Q}$ = (0 0.27 6.5). Solid lines are fitted functions of the corresponding elements. Red line is the sum of elastic lines and phonons

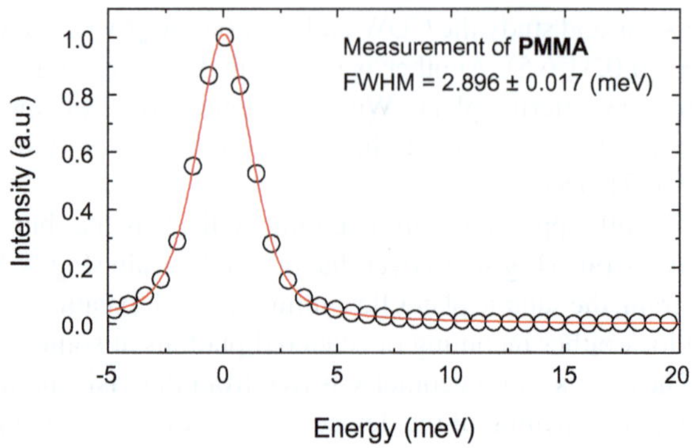

Fig. 4.5 Inelastic spectrum measured from PMMA. Fitting using the pseudo-Voigt function (red line) gives a perfect fit with the FWHM (Full Width Half Maximum) of 2.963 ± 0.017 (meV)

The study of the CDW peak width and intensity against external perturbations, such as magnetic field, temperature or uniaxial strain, is one of the typical methods to understand the CDW phenomenology in cuprates.

4.2.2 Energy Resolution

The energy resolution of the IXS spectrometer depends on many factors, such as the Darwin width of the crystal monochromator, geometrical factors from the optics and any imperfection of the x-ray optics. Our experiment used an incident photon energy of 17.794 keV from the Si(9 9 9) reflection of the crystal monochromator. The total energy resolution of the experimental setup was measured using PMMA as a scatterer, as shown in Fig. 4.5. It gives an energy resolution of 2.963 ± 0.017 meV consistent with the expectations for ID-28.[2]

4.2.3 Momentum Resolution

The spectrometer is 7 m away from the sample and the detector slit size was set to 20 (horizontal) × 50 (vertical) (mm^2) during the measurement. Since the beam size ($50 \times 40\ \mu m^2$) is much smaller than the detector slit and the beam divergence of 40 (horizontal), 20 (vertical) μrad is negligibly small compared to the solid angle of the detector (20/7000 ∼ 2.86 mrad) with respect to the sample position, only the geometric path of the beam is taken into account here.

With this information, our momentum resolution can be estimated in the following way. Using the Laue condition ($\Delta \boldsymbol{k} = \boldsymbol{Q}$) and elementary trigonometry, we have

[2] https://www.esrf.eu/UsersAndScience/Experiments/DynExtrCond/ID28/Characteristics.

$$|Q|^2 = 2|k|^2(1 - cos(2\theta))$$
$$|Q| = \sqrt{2}|k|\sqrt{1 - cos(2\theta)}$$
$$\Delta|Q| = \sqrt{2}|k|\frac{sin(2\theta)}{\sqrt{1 - cos(2\theta)}}\Delta(2\theta)$$

Our measurements near $\boldsymbol{Q} = (0\ 0.31\ 7)$ give $2\theta \sim 24°$. Plugging $2\theta = 24°$ and $\Delta(2\theta) = 20/7000$ rad into the above equation gives $\Delta|Q| = 0.004$ (1/Å). Projection of $\Delta|Q|$ onto K and L results in momentum resolutions of $\Delta K \sim 0.0016$ (r.l.u.) and $\Delta L \sim 0.023$ (r.l.u.) in HWHM.

4.3 Results

4.3.1 Enhancement of 2D-CDW Under Uniaxial Pressure

Figure 4.6 compares the IXS intensity measured without strain and at 1.0% compression along the a-axis. Several spectra were measured at various momentum transfers around $\boldsymbol{Q} = (0\ 0.31\ 6.5)$ along K. The spectral features observed in the accessible

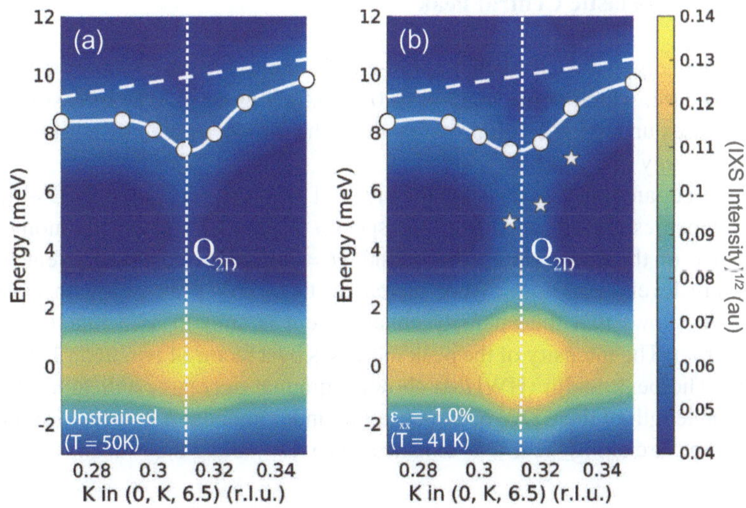

Fig. 4.6 Color plot of the raw IXS intensity as a function of momentum transfer $\boldsymbol{Q} = (0\ \text{K}\ 6.5)$ along K. For better visualization with higher contrast, square root of the IXS intensity was plotted. Dashed lines represent the calculated dispersion of the low-lying acoustic mode. The solid lines with white points are a guide to the eye that shows a superconductivity-induced softening. Same measurements were done **a** in the unstrained condition and **b** under a compression of −1.0%. Stars indicate the spectral weight from the soft optical phonon mode. Vertical dotted line shows the position of the CDW wave vector in the reciprocal space. From [14]. Reprinted with permission from AAAS

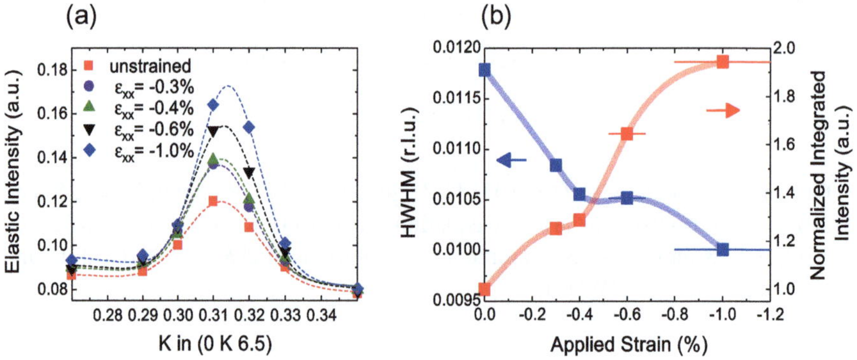

Fig. 4.7 **a** Elastic intensity scans of 2D-CDW as a function of uniaxial pressure. **b** Integrated intensity, peak width HWHM against uniaxial pressure. From [14]. Reprinted with permission from AAAS

energy range can be categorized into two main contributions: quasi-elastic intensity (at zero energy) and phonons (finite energy).

4.3.1.1 Quasi-elastic Central Peak

Near zero energy, a well-known quasi-elastic central peak is observed in the unstrained case, which peaks at $\boldsymbol{Q} = (0\ 0.31\ 6.5)$ (Fig. 4.6a). The intensity of this peak increases under 1.0% compression along the a-axis, as seen by a brighter spot near zero energy in Fig. 4.6b.

To give a quantitative analysis on the uniaxial pressure effect on the CDW, quasi-elastic intensities are extracted from the spectra. Figure 4.7a shows the momentum dependence of the elastic intensity around $\boldsymbol{Q} = (0\ 0.315\ 6.5)$ at different uniaxial pressure levels. As the pressure increases, the CDW peak grows in intensity. Figure 4.7b shows that the intensity enhancement is smooth as a function of the applied strain. The intensity of the peak increases by a factor of two upon 1.0% compression. The peak width (HWHM) shows a gradual decrease, indicating that the correlation length of the CDW becomes larger under uniaxial pressure. In summary, uniaxial pressure along the a-axis enhances the CDW along the b-axis.

4.3.1.2 Phonon

At excitation energies around 8 meV, the dispersion of the low-lying acoustic phonon (solid white line) can be observed for the unstrained case in Fig. 4.6a. The Kohn anomaly at the CDW wave vector $\boldsymbol{Q}_{2D} = (0\ 0.31\ 6.5)$ discovered in the previous study [1] is nicely reproduced here. The calculated phonon dispersion from DFT (white dashed lines) does not capture the observed phonon softening, indicating that it stems from some strong correlation effect.

At 1.0% compression along the *a*-axis (Fig. 4.6b), the Kohn anomaly of the low-lying acoustic mode is still visible. On top of that, there is an additional spectral weight below the energy of the acoustic phonon mode, makred by stars. This is a distinct new feature which is not observed in the unstrained case. The origin of the new low-energy spectral weight will be discussed in the following section.

4.3.2 *Emergence of 3D-CDW Under Uniaxial Pressure*

4.3.2.1 In-Phase Correlation Along the *c*-axis

Figure 4.8a shows the L-dependence of the elastic intensity at different strain levels. In the unstrained case, we observed the well-known broad peak centered around L $\sim$ 6.5, typical of the quasi-2D CDW at ambient pressure in YBCO. The inverse of the peak width along L (FWHM $\sim$ 0.5 r.l.u.) provides the correlation length along the *c*-direction, which gives a rough estimate of less than one unit cell ($\xi_c = \frac{c}{\pi \times FWHM} \sim 0.6c$) along the *c*-direction. This indicates that the correlation does not even extend beyond one unit cell, mainly showing a 2D-character. It is weakly peaked at an half integer L value ($L_{peak} = 6.5$), which indicates that it has a very weak anti-correlation between neighboring unit cells.

As uniaxial pressure along the *a*-axis increases, a small but growing feature begins to appear at L = 7. At the uniaxial pressure of 1.0% along the *a*-axis, a sharp, intense peak appears at L = 7. The sharpness of the peak, compared to the broad bump of

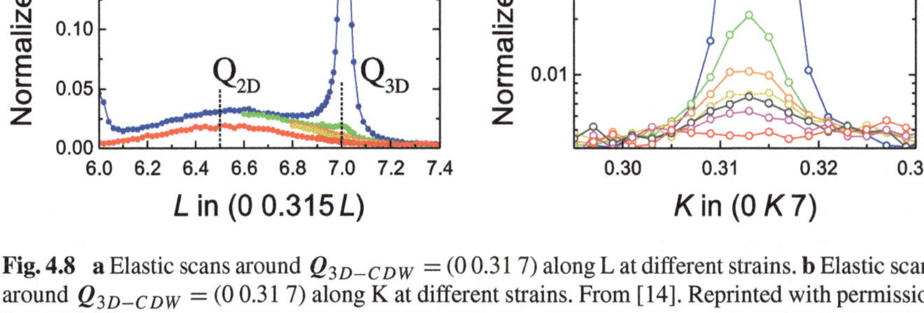

Fig. 4.8 **a** Elastic scans around $Q_{3D-CDW} = (0\,0.31\,7)$ along L at different strains. **b** Elastic scans around $Q_{3D-CDW} = (0\,0.31\,7)$ along K at different strains. From [14]. Reprinted with permission from AAAS

the 2D-CDW, indicates a long correlation length along the c-axis, revealing a three-dimensional character of the new charge modulation. The integer value of the peak center shows an in-phase correlation along the c-axis, distinct from the 2D-CDW with the weak anti-correlation between unit cells.

Figure 4.8b shows the strain dependence of the elastic peak along K in the reciprocal space. The new 3D-CDW peak is centered at K $\sim$ 0.315, showing the same incommensurability as the 2D counterpart. Note that the plot uses a logarithmic scale, in order to show both the sudden appearance of the 3D-CDW intensity upon a-axis compression and its much stronger intensity compared to the 2D-CDW.

From the elastic scans along K and L, the HWHMs of the peak are estimated to be $\sigma_b \sim 0.002$ r.l.u. and $\sigma_c \sim 0.02$ r.l.u. Since these values are comparable to the momentum resolution limit ($\Delta K \sim 0.0016$ r.l.u. and $\Delta L \sim 0.023$ r.l.u.), it is reasonable to take the measured inverse peak widths as lower bounds of the correlation lengths.

Note that a new peak appears at L = 6 as well. The presence of the new peaks at integer L (L = 6, 7) indicates that the new peak arises from the periodic arrangement of charges that is commensurate with the underlying lattice along the c-axis. This, in turn, rules out the possibility that this has an extrinsic origin (not related to the bulk sample under strain), e.g. a Bragg peak of a broken piece of YBCO resulting from the high level of strain.

4.3.2.2 Long-Range Correlation of 3D-CDW

The peak widths along K and L give correlation lengths of $\xi_b \sim 80\text{b} \sim 310$ Å and $\xi_c \sim 8\text{c} \sim 94$ Å. Elastic scans along H were not possible due to limitations in the scattering geometry, where the H-axis is perpendicular to the scattering plane. Although a direct measurement of the correlation length of the 3D-CDW along H-axis will be possible only with a proper four-circle diffractometer, we take the correlation length along the a-axis to be $\xi_a \sim \xi_b \sim 310$ Å, following the case of the 2D-CDW, where the correlation lengths along the a and b-axis are similar. This provides a correlation volume of the 3D-CDW as $\xi_a \times \xi_b \times \xi_c \sim 9033400\,\text{Å}^3 \sim 52000$ unit cells. This value is much greater than the correlation volume of 2D-CDW nanodomains ($\sim$250 unit cells), demonstrating the distinctive long-range nature of the uniaxial-strain-induced 3D-CDW.

4.3.2.3 Competition with Superconductivity

Figure 4.10 shows the temperature dependence of the strain-induced 3D-CDW quasi-elastic peak. A clear peak at L = 7 is visible at T = 70 K, well above the superconducting critical temperature T_c at ambient pressure (T_c = 65 K). This is the first observation of a 3D-CDW without the suppression of superconductivity by a high magnetic field. Our finding of the 3D-CDW above the superconducting dome implies that the high field state of the cuprates is not a necessary condition for the formation

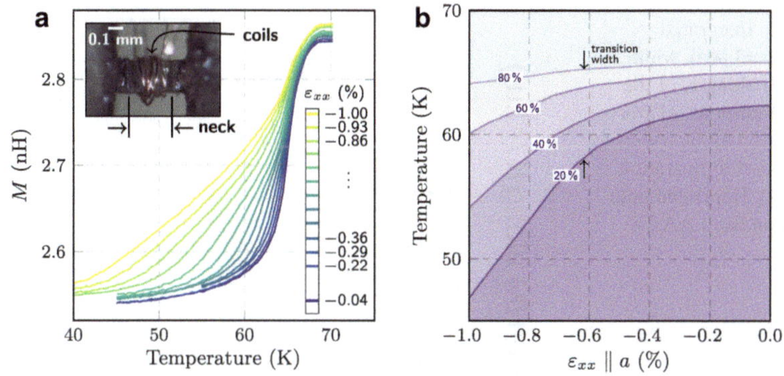

Fig. 4.9 **a** Mutual inductance measurement at different strain levels along a-axis. The inset shows the photo of the coils wrapped around the sample under strain. **b** Constant mutual inductance levels extracted from the data in panel **a**. 20, 40, 60 and 80% lines correspond to the cuts at 2.60, 2.66, 2.72, and 2.78 nH. From [14]. Reprinted with permission from AAAS

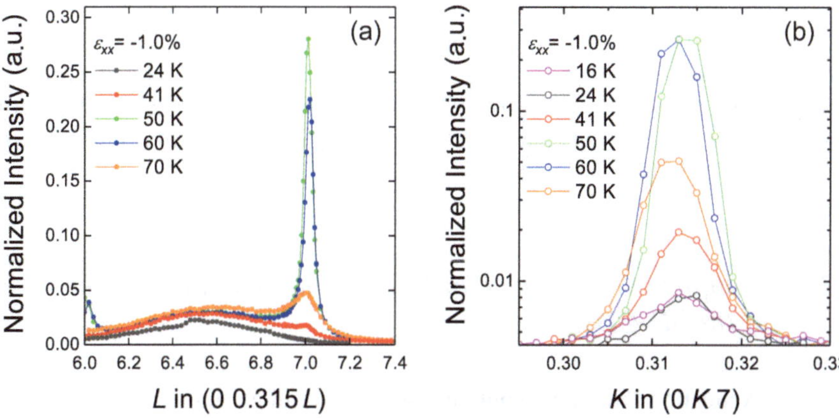

Fig. 4.10 **a** Elastic scans around $\boldsymbol{Q}_{3D-CDW} = (0\,0.31\,7)$ along L at different temperatures. **b** Elastic scans around $\boldsymbol{Q}_{3D-CDW} = (0\,0.31\,7)$ along K at different temperatures. From [14]. Reprinted with permission from AAAS

of the 3D-CDW. This implies that the 3D-CDW can grow without the aid of the superconducting vortex cores as nucleation centers.

The intensity of the 3D-CDW peak is maximum around T = 50–60 K. Below T = 50 K, it loses most of its intensity and at T = 24 K, it is hardly visible. The overall behavior of the 3D-CDW elastic peak against temperature is visualized in Fig. 4.11. The estimated superconducting T_c at -1.0% a-axis compression is around 50–60 K (Fig. 4.9). This coincides with the temperature, where the 3D-CDW peak starts to drop in intensity, implying that like the quasi-2D CDW at ambient pressure, the strain-induced 3D-CDW exhibits a clear competition with superconductivity. It is worth noting that the 3D-CDW shows a stronger competition with superconductivity

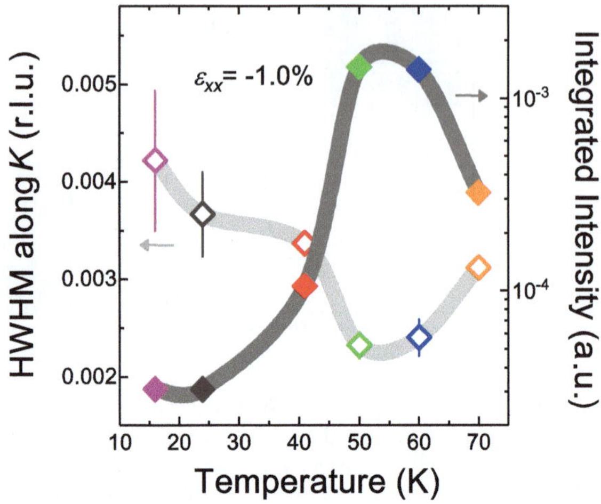

Fig. 4.11 Integrated intensity and peak width (HWHM) along K of the 3D-CDW peak at −1.0% compression along *a*-axis as a function of temperature. From [14]. Reprinted with permission from AAAS

than the quasi-2D CDW. This is evident from Fig. 4.10a. The 2D-CDW broad peak centered at L = 6.5 decreases by half in intensity at T = 24 K, compared to its maximum intensity at T = 50–70 K, whereas the intensity of the 3D-CDW sharp peak at L = 7 decreases dramatically by more than an order of magnitude, when going from T = 50 K to T = 24 K.

4.3.3 *Complete Softening of an Optical Phonon at the Onset of the 3D-CDW*

4.3.3.1 Acoustic Phonon Anomaly in the Absence of Strain

Figure 4.12a shows inelastic spectra measured at different K along (0 K 7) in the reciprocal space in unstrained conditions. In the energy range [0, 20] meV, three[3] distinct phonon peaks are visible. Similarly to the phonon softening observed for the 2D-CDW, a Kohn anomaly is observed in a narrow region around the 3D-CDW wave vector $\boldsymbol{Q}_{3D}$ = (0 0.31 7), as marked by gray arrows. Away from $\boldsymbol{Q}_{3D}$, the observed IXS intensity follows the calculated phonon dispersion, depicted by red lines.

4.3.3.2 Softening of a New Optical Phonon Under Strain

At −1.0% strain, where the quasi-elastic peak of the 3D-CDW is observed, the acoustic phonon anomaly is no longer visible. In Fig. 4.12b, the gray arrows show

[3] Actually, two phonon branches exist around 10 meV, which cannot be resolved within our experimental energy resolution.

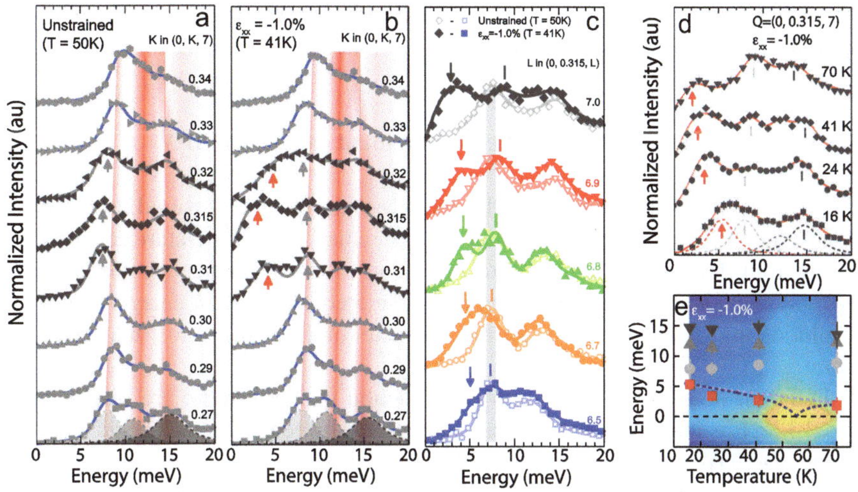

Fig. 4.12 **a** IXS intensity measured at different K along (0 K 7) in the absence of strain. Solid lines represent the fitted result, each phonon's contribution of which is shown on the spectrum measured at K = 0.27. **b** Same, measured at −1.0% compression along *a*-axis and T = 41 K, to remain below $T_c(\epsilon_{xx} = -1.0\%) \sim$ 50–60 K. Red lines on **a** and **b** indicate the calculated dynamic structure factor of the phonons. Gray arrows indicate a low-lying acoustic phonon. Red arrows point to a new soft optical phonon. **c** Comparison of the inelastic spectra between the unstrained state and −1.0% compression along *a*-axis, measured at different L along (0 0.315 L) in the reciprocal space. For the spectra under *a*-axis compression, arrows indicate the energy of the soft optical phonon mode, whereas ticks indicate the energy of the acoustic phonon mode. The long gray column represents the energy of the acoustic phonon without strain. **d** Inelastic spectra measured at $\boldsymbol{Q}$ = (0 0.315 7) as a function of temperature. **e** Color plot of the IXS intensity as a function of temperature on a log scale. Data points correspond to the energy of each phonon observed in the spectra. Dotted lines are guides to the eye that represent possible scenarios of the temperature dependence of the soft optical phonon. From [14]. Reprinted with permission from AAAS

that the energy of the acoustic phonon does not soften, but agrees rather reasonably with the calculated phonon dispersion. Strikingly, another phonon (marked by red arrows) undergoes a remarkable softening. This has not been observed in previous measurements and is not predicted by the DFT calculation. The energy of the new phonon is even below the existing acoustic phonon energy and we assign it to an optical phonon. The softening of the new optical phonon is only observed in a narrow region of the reciprocal space near $\boldsymbol{Q}_{3\mathrm{D}}$ = (0 0.315 7) along K and gets stronger as it approaches the 3D-CDW wave vector $\boldsymbol{Q}_{3\mathrm{D}}$ along K.

The magnitude of the optical phonon softening is also observed to be maximum at $\boldsymbol{Q}_{3\mathrm{D}}$ when scanned along L, as shown in Fig. 4.12c. The low-energy optical phonon, as shown by the arrows, is visible at all L values L = 6.5–7 under −1.0% strain and it softens more as it approaches L = 7, i.e. $\boldsymbol{Q}_{3\mathrm{D}}$. The coincidence of the momentum dependence of the optical phonon softening with $\boldsymbol{Q}_{3\mathrm{D}}$ indicates an intimate relation between the optical phonon softening and the emergence of the 3D-CDW. It is interesting to note that under 1.0% compression, the acoustic phonon hardens as it approaches $\boldsymbol{Q}_{3D}$, showing anti-correlation with the optical phonon softening.

4.3.3.3 Soft Phonon Mode in the Development of the 3D-CDW

Figure 4.12d shows the temperature dependence of the spectra measured at $\boldsymbol{Q}_{3D}$ under -1.0% strain. The new optical phonon shows a remarkable temperature dependence. Going from 70 to 16 K in temperature, the energy of the new optical phonon changes by more than 3 meV, whereas other phonons merely fluctuate in energy within the resolution limit.[4] This reveals a distinct character of the new phonon induced under strain.

Different responses of phonons as a function of temperature are displayed in Fig. 4.12e. It is interesting to note the correlation between the quasi-elastic intensity of the 3D-CDW and the new optical phonon (red square) energy. As the 3D-CDW approaches the maximum intensity (bright spot near T = 50–60 K), the optical phonon softens close to zero energy. Below T = 50 K in the superconducting state, the 3D-CDW intensity decreases dramatically and the optical phonon hardens. This is reminiscent of the soft-phonon-mode-driven CDW observed in 2H-$NbSe_2$ (also marked as black-dotted lines in Fig. 4.12e).[5]

4.4 Discussion

4.4.1 Comparison with Magnetic-Field-Induced 3D-CDW

Our uniaxial pressure study is not the first discovery of 3D-CDW order in underdoped YBCO. Recently, several x-ray scattering studies [15, 16] on YBCO under high magnetic field (close to or larger than H_{c2}) observed the emergence of the 3D-CDW. The magnetic-field-induced 3D-CDW has the same incommensurability as the 2D-CDW and shows distinct features in thermodynamic probes, such as sound velocity measurements [17], Hall effect measurements [18] and NMR [19].

It is not directly self-evident whether the magnetic-field-induced 3D-CDW and the uniaxial-pressure-induced 3D-CDW share the same origin. Experimentally, they do share the same phenomenology, such as the 3D-CDW having the same incommensurability as its 2D counterpart, much larger correlation volume than that of the 2D-CDW, and a preceding 2D-CDW enhancement prior to the 3D-CDW formation. The suppression of superconductivity is in favor of the formation of the 3D-CDW in both cases. In the case of magnetic field, the 3D-CDW was observed only along the *b*-

[4] Presence of multiple phonons between 8 and 20 meV and the experimental energy resolution of ~3 meV make it difficult to extract the exact phonon energy.

[5] However, the scenario of a simple hardening upon cooling marked as gray dotted lines cannot be excluded.

axis and not along the a-axis. Interestingly, the uniaxial-pressure-induced 3D-CDW was also observed along the b-axis. Whether uniaxial pressure can induce 3D-CDW along the a-axis is an interesting question that will be addressed in the next chapter.

4.4.2 Comparison Between 2D- and Uniaxial-Pressure-Induced 3D-CDW

The quasi 2D-CDW observed in YBCO and all cuprate families is generally not in line with the conventional charge density wave systems, which is qualitatively described by the Peierls theory. The theory of Peierls predicts that an electronic instability arises in low-dimensional metals due to the divergent charge susceptibility at the nesting vector connecting two points on the Fermi surface. Finding a nesting vector on the Fermi surface of underdoped YBCO can test the validity of this approach. Unfortunately, YBCO is not the best material to study with ARPES. It tends to be cleaved between the BaO layer and the CuO chain layer, not on the CuO_2 plane [20, 21]. Moreover, the top-most layer of YBCO becomes heavily overdoped [22] after cleaving, preventing a study of the bulk properties of underdoped YBCO.

Apart from the absence of reliable Fermi surface information, there are many experimental features that are not easily understood in the framework of conventional CDW systems: (i) the 2D-CDW in YBCO has a broad transition, compared to sharp CDW transitions in systems like 2H-NbSe2 [23] and $ZrTe_3$ [24]. (ii) Experimental evidence of the 2D-CDW phase from thermodynamic probes is still lacking. (iii) Electron-phonon coupling does not lead to a complete softening of the phonon, even when the quasi-elastic CDW peak is established. (iv) Phonon softening occurs at the superconducting transition temperature, not at the CDW transition temperature.

On the other hand, the uniaxial-pressure-induced 3D-CDW shows many features that are opposite to the features of the 2D-CDW: (i) It has a sharp transition with respect to uniaxial pressure and temperature. (ii) Since the discovery of the 3D-CDW under strain is new, there have not yet been studies with thermodynamic probes under sufficiently high strain. However, various studies [17–19] have detected a thermodynamic transition under high magnetic field that coincides with the 3D-CDW transition under magnetic field. (iii) An apparent complete softening of an optical phonon due to strong electron-phonon coupling at the 3D-CDW wave vector coincides with the emergence of the induced 3D-CDW peak.

4.5 Conclusions

The CDW in YBCO was studied using non-resonant inelastic x-ray scattering under uniaxial pressure applied along the a-axis, which makes the lattice more orthorhombic. a-axis pressure not only enhances the existing quasi-2D CDW along the b-axis, but also develops a new 3D-CDW that is commensurate with the c-axis periodicity. The correlation volume of the 3D-CDW is two orders of magnitude larger than that of the 2D-CDW. It is remarkably long-ranged, unlike the 2D-CDW nano-domains.

The complete softening of the new optical phonon, which is not present in the unstrained case, is observed, happening alongside the onset of the 3D-CDW. This fits well into the conceptual framework of soft-phonon-mediated structural phase transition, consistent with the notion that the uniaxial-pressure-induced 3D-CDW is a thermodynamic state in YBCO.

Our research opens a new research field of studying charge density waves using uniaxial strain, investigating a previously unexplored phase space of high-T_c cuprates. It also emphasizes again the importance of strong electron-phonon interactions in the formation of CDWs in YBCO.

References

1. Le Tacon M et al (2014) Nat Phys 10:52–58
2. Blackburn E et al (2013) Phys Rev B 88:054506. https://doi.org/10.1103/PhysRevB.88.054506. Accessed 5 Aug 2013
3. Jorgensen J et al (1990) Phys C: Supercond 171:93–102. https://doi.org/10.1016/0921-4534(90)90460-V
4. Struzhkin VV, Chen X-J (2016) Low Temp Phys 42:884–890. https://doi.org/10.1063/1.4965891
5. Sadewasser S et al (2000) Phys Rev B 61, PRB:741–749
6. Souliou SM et al (2018) Phys Rev B 97:020503. https://doi.org/10.1103/PhysRevB.97.020503. Accessed 2 Jan 2018
7. Huang H et al (2018) Phys Rev B 97:174508. https://doi.org/10.1103/PhysRevB.97.174508. Accessed 17 May 2018
8. Putzke C et al (2018) Phys Rev Lett 120, PRL:117002(2018)
9. Cyr-Choinière O et al (2018) Phys Rev B 98, PRB:064513
10. Vinograd I et al (2019) Phys Rev B 100:094502. https://doi.org/10.1103/PhysRevB.100.094502. Accessed 9 Sept 2019
11. Welp U et al (1992) Phys Rev Lett 69, PRL:2130–2133
12. Hicks CW et al (2014) Science 344:283–285. https://doi.org/10.1126/science.1248292. https://science.sciencemag.org/content/344/6181/283.full.pdf
13. Lei M et al (1993) Phys Rev B 47:6154–6156. https://doi.org/10.1103/PhysRevB.47.6154. Accessed 10 March 1993
14. Kim H-H et al (2018) Science 362:1040–1044. https://doi.org/10.1126/science.aat4708. https://science.sciencemag.org/content/362/6418/1040.full.pdf
15. Gerber S et al (2015) Science 350:949–952. https://doi.org/10.1126/science.aac6257. https://science.sciencemag.org/content/350/6263/949.full.pdf
16. Chang J et al (2016) Nat Commun 7:11494. https://doi.org/10.1038/ncomms11494
17. LeBoeuf D et al (2013) Nat Phys 9, 093FL Times Cited:137 Cited References Count:27:79–83
18. LeBoeuf D et al (2007) Nature 450:533
19. Wu T et al (2011) Nature 477, 816MC Times Cited:366 Cited References Count:30:191–194
20. Derro DJ et al (2002) Phys Rev Lett 88:097002. https://doi.org/10.1103/PhysRevLett.88.097002. Accessed 9 Feb 2002
21. Hossain MA et al (2008) Nat Phys 4:527

22. Zabolotnyy VB et al (2007) Phys Rev B 76:064519. https://doi.org/10.1103/PhysRevB.76.064519. Accessed 6 Aug 2007
23. Moncton DE, Axe JD, DiSalvo FJ (1977) Phys Rev B 16, PRB:801–819
24. Hoesch M et al (2009) Phys Rev Lett 102, PRL:086402

Chapter 5
CDW in $YBa_2Cu_3O_{6.67}$ Under Uniaxial Pressure: Resonant Inelastic X-ray Scattering

5.1 Background

Previous inelastic x-ray scattering experiments on CDW under uniaxial pressure provided a new way of realizing the 3D-CDW in cuprates at -1.0% strain along the a-axis, without the need of a magnetic field exceeding 15 T as in previous work on 3D-CDW. The technical challenges connected to the implementation of uniaxial strain are minor compared to those of the experiments under high fields, thereby giving uniaxial strain a clear advantage for the extensive investigation of the 3D-CDW using a variety of experimental techniques.

Although inelastic x-ray scattering experiments provided us detailed information regarding the intimate relation between charge density waves and phonons, its photon-hungry nature along with the limited accessible geometries make it challenging to perform comprehensive studies exploring large volumes in reciprocal space. Taking momentum-dependent spectra along one principal axis with good statistics required almost 10 h in our previous measurements. Considering that typical x-ray diffraction experiments require measurement times ranging from a few seconds to minutes for one scan, IXS is extremely time-consuming and slow. In addition, once a scattering plane has been defined, for example (0 K L), the IXS setup and cryostat do not allow one to change the scattering plane to (H 0 L), without removing the sample and remounting it.

Specific to underdoped $YBa_2Cu_3O_{6+x}$, the oxygen order in the chain layer makes it difficult to study the CDW along the a-axis, as mentioned in the previous chapter. Our sample of interest has a doping level $p = 0.12$, where the ortho-VIII oxygen order is present. This gives a substantial elastic intensity at the wave vector $\boldsymbol{Q}_{//} =$ (0.375 0) in 2D-Brillouin Zone (2D-BZ). Thus, very weak CDW signals are easily shadowed even by the tails of the chain elastic peaks.

We chose to perform resonant x-ray scattering (RXS) measurements to overcome this issue. RXS at the Cu-L_3 edge has been extensively used in the research of the CDW in cuprates. By tuning the energy to the resonant edge of Cu(2) atoms in the

H. Kim, *Uniaxial Pressure Study of Charge Density Waves in a High-T_c Cuprate Superconductor*, Springer Theses,
https://doi.org/10.1007/978-3-030-99898-1_5

CuO_2 planes (~931.3 eV), it greatly enhances the scattering cross section of the CDW signal upon others. This significantly reduces the problem of having intense elastic background signals from the chain order and enables us to access all geometries, including those to study the CDW along the *a*-axis.

5.2 Experiment

Resonant inelastic x-ray scattering (RIXS), which was described in Chap. 3, combines the energy resolving capabilities of IXS to single-out the quasi-elastic scattering signal, with the advantages of resonant enhancement of the scattering from Cu atoms in the CuO_2 planes while excluding the signal from the chains. RIXS experiments were carried out on the ERIXS spectrometer at the beamline of ID32 in the ESRF. The utilization of the in-vacuum four-circle diffractometer manipulator of ERIXS removes the limitations in the scattering geometry of the previous experiment. It provides access to new directions in reciprocal space, covering the wave vectors of charge density waves both along the *a*- and *b*-axis in a single experiment.

Figure 5.1a shows the scattering geometry of the experiment with the strain device. The sample is placed at the center of rotation, so that it stays on the incident beam regardless of changes in angles. In order to maximize the scattering signal from charge, σ-polarization of the incident beam was used. The scattered photons are not polarization-resolved, therefore contain both σ- and π-polarizations.

With the scattering geometry in Fig. 5.1a, a CDW whose modulation is perpendicular to the direction of uniaxial pressure can be studied. A study of the CDW parallel to the uniaxial pressure direction can be performed by rotating the sample by 90° around the *c*-axis (ϕ-rotation). This is illustrated in Fig. 5.1b. With this geometry, the finite thickness of the sample plate restricts the scattering angle of the experiment and has a potential risk of beam shadowing by the sample plates. Our experiment was designed for an angular opening of 132° in this geometry. The absence of shadowing

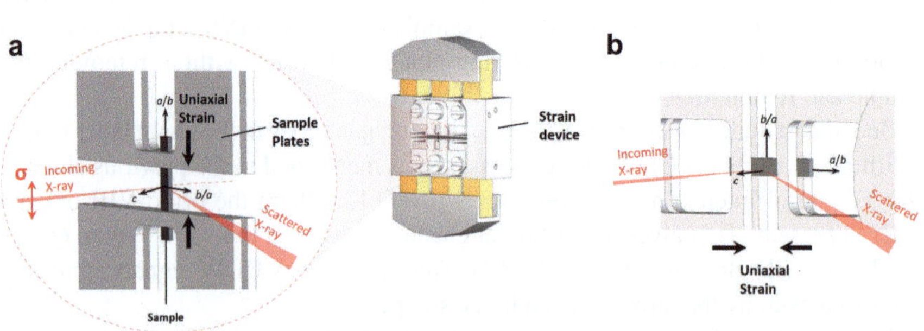

Fig. 5.1 Schematic picture of the scattering geometry with the strain device, when CDW of interest is **a** perpendicular to the direction of uniaxial pressure **b** parallel to the direction of uniaxial pressure. Reprinted figure with permission from [1]. Copyright (2021) by the American Physical Society

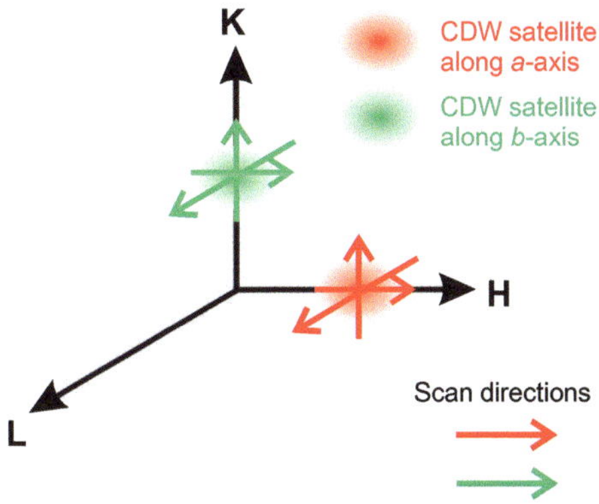

Fig. 5.2 Schematic picture of the momentum scans in the experiment. For each of the CDW satellites along the *a*- and *b*-axis, scans along H, K and L were done

in this range was carefully checked by monitoring the intensity of *dd*-excitations during the scan, which is expected to remain constant in σ-polarization.

The goal of the experiment is to collect comprehensive information on the CDW responses under uniaxial pressure. This involves all combinations of *a*-axis/*b*-axis pressure and CDW wavevector along the *a*-axis/*b*-axis. Since changing the strain direction in a single experiment is technically not possible, we took the following approach.

First, under uniaxial pressure along the *a*-axis, we did measurements on the CDW along the *a*- and *b*-axes. Then, we switched to the experiment under uniaxial pressure along the *b*-axis and collected the information on the CDW along the *a*- and *b*-axes again. For each measurement on the CDW satellite peaks, momentum cuts along H, K and L were scanned, as shown in Fig. 5.2. Some of the scans were not performed due to the angular limit of the motors or beam shadowing. For example, the scans along L involves a grazing condition for either the incident or scattered beam. This becomes problematic when measuring an L-scan of the CDW parallel to the pressure direction (Fig. 5.1), since one of the beams is completely shadowed by the sample plates.

5.2.1 Typical RIXS Spectrum

A typical RIXS spectrum of underdoped YBCO is shown in Fig. 5.3. Various features are observed in different energy ranges.

(i) Elastic signal: The intensity comes from the photons elastically scattered from the sample. It is observed as a sharp peak at zero energy loss. It is not a delta-function because of the instrumental factor (e.g. finite bandwidth from the grating of the monochromator, the grating of the spectrometer, and the detector). In our measurements, we chose a combined resolution of the setup ~60 meV, which is coarser than the highest resolution offered by the spectrometer, but

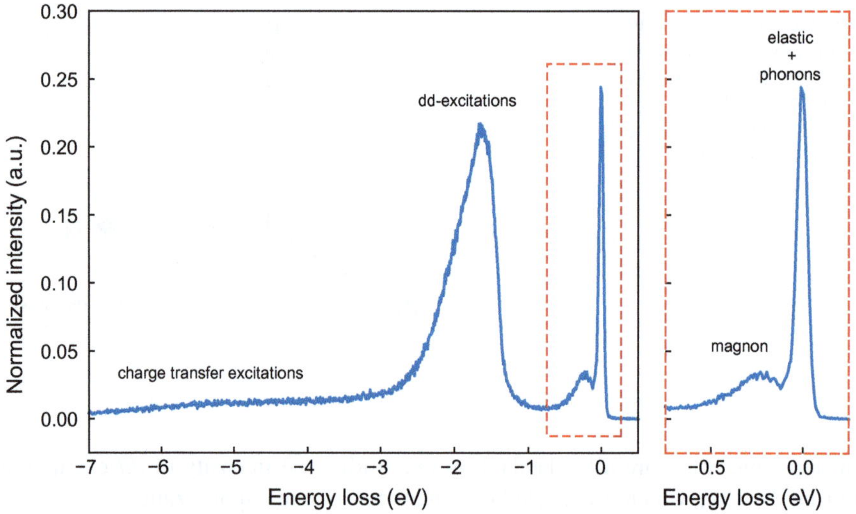

Fig. 5.3 Example RIXS spectrum of $YBa_2Cu_3O_{6.67}$ measured at $\boldsymbol{Q}_{//} = (0\ 0.315)$ with incident π-polarization. The plot on the right side in the red rectangle is the enlarged spectrum in the energe range [−0.75 0.25]. Reprinted figure with permission from [1]. Copyright (2021) by the American Physical Society

allows for a reasonable photon flux. On top of a background coming from the tail of the Fresnel reflectivity, the residual roughness of the sample surface and thermal diffuse scattering, the intensity of the elastic peak represents the quasi-static electronic charge density with the measured wave vector. Therefore, the momentum dependence of this intensity is used to identify the CDW modulation, as was done in the previous chapter. The analysis of the elastic intensity under uniaxial pressure will be the main content of this chapter.

(ii) Phonons: They lie in the energy range from 0 to ~80 meV. Since our energy resolution is comparable to the phonon energy, it is quite challenging to accurately separate the weaker phonon intensities from the stronger elastic intensity. This is different from the non-resonant inelastic x-ray scattering experiment in the previous chapter, whose high energy resolution allowed tracking of individual phonon dispersions. Therefore, analysis of the phonons is not a main goal of the RIXS study.

(iii) Magnons: They disperse as a function of momentum in the energy range ~0–300 meV. Since it is quite separated from the elastic peak, thanks to the large superexchange coupling J in cuprates, one can extract the detailed spectroscopic information from the spectra. Magnon peaks are observed to be sharper as the system goes towards lower doping levels and get more damped at higher doping. The magnons did not show a noticeable change under uniaxial pressure and will not be discussed in this chapter.

(iv) *dd*-excitations: They occupy the energy range ~1–3 eV. These are orbital excitations of the $3d^9$ hole from $d_{x^2-y^2}$ to other d-orbitals, such as $d_{xy,yz,zx}$, $d_{3z^2-r^2}$ and they are usually the most distinctive feature in the spectrum. The *dd* signal shows distinctive peaks for each excitation as the doping level decreases, whereas they get broader as the hole doping increases. No big change in the *dd*-excitations was observed under uniaxial pressure. These excitations are not treated in this chapter.

(v) Charge transfer excitations: They give rise to a very broad feature observed in the energy range ~3–10 eV. When an incident photon excites an electron from the Cu-$2p$ to the Cu-$3d$ state, it creates a core hole in the Cu-$2p$ state. The strong core-hole Coulomb potential causes charge from the ligand oxygen to move into the Cu site to screen the core hole. Then, the excited $3d$ electron decays to fill the $2p$ core hole, leaving an electron-hole charge-transfer excitation in the final state. Therefore, the energy of the scattered photon corresponds to the energy of charge transfer excitations from the oxygen ligands to the Cu-$3d$ states.

5.2.2 Energy Resolution

The energy resolution of the RIXS spectrometer depends on various factors. The bandwidth of the incident x-ray beam, which is determined by the grating of the monochromator and the opening of the entrance slit, directly affects the combined energy resolution of the beamline and the spectrometer. The slope error and roughness of the variable line spacing spherical grating used in the ERIXS analyzer also influence the energy resolution. Other contributions include the beam spot size at the sample and the detector. The combined energy resolution can be obtained by measuring a spectrum of a smooth surface in specular conditions. Figure 5.4 shows an elastic spectrum measured on a multilayer of $[W/B_4C]\times 10$. The FWHM of the elastic peak is 3.3 pixels, which gives ~60 meV when converted to energy. All the spectra were measured with this energy resolution throughout the experiment.

5.2.3 Momentum Resolution

In our RIXS setup, the momentum resolution is mainly limited by the angular acceptance of the horizontal collimating mirror, which is located 1.1 m away from the sample with an angular acceptance of 20 mrad. The schematic picture of the optics after the sample is visualized in Fig. 5.5. As was done in the previous chapter, using the Laue condition ($\Delta \boldsymbol{k} = \boldsymbol{Q}$) and elementary trigonometry, we have

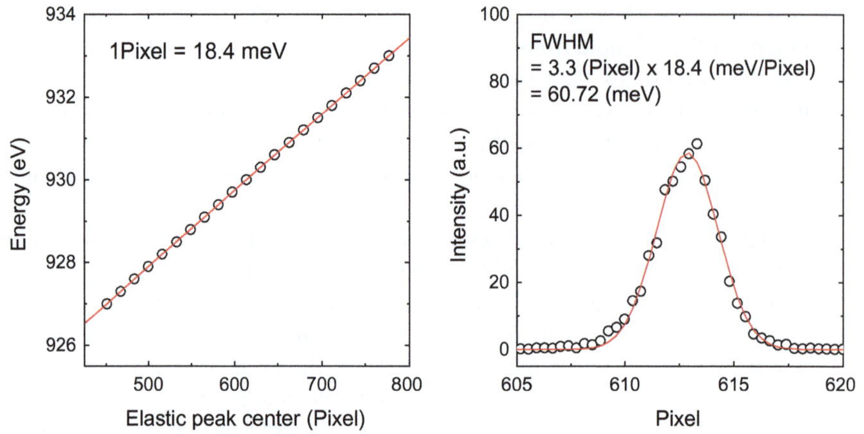

Fig. 5.4 (left) Center of the elastic peak position on the CCD as a function of the incident x-ray energy. (right) Elastic peak in the spectrum of W/B_4C

Fig. 5.5 Brief scheme of the RIXS grating spectrometer. Reprinted from [2], with the permission of AIP Publishing

$$|Q|^2 = 2|k|^2(1 - cos(2\theta))$$
$$|Q| = \sqrt{2}|k|\sqrt{1 - cos(2\theta)}$$
$$\Delta|Q| = \sqrt{2}|k|\frac{sin(2\theta)}{\sqrt{1 - cos(2\theta)}}\Delta(2\theta)$$

Using the $2\theta = 149.5°$ and $\Delta(2\theta) = 20\,$mrad gives $\Delta|Q| = 4\times\ 10^{-4}$ (1/Å). For $\boldsymbol{Q}_{//} = (0\,0.31)$, the projection of $\Delta|Q|$ on the 2D-Brillouin Zone results in a momentum resolution of $\Delta|Q|_{//} = 0.001$ (r.l.u.). It is an order of magnitude smaller than the typical peak width of the CDW (HWHM $\sim$ 0.01 (r.l.u.)), so we neglect the contribution from the momentum resolution in our analysis.

5.3 Results

For simplicity, the 2D-CDW with the modulation vectors $\boldsymbol{Q}^a_{//} = (0.305\,0)$ and $\boldsymbol{Q}^b_{//} = (0\,0.315)$ will be called hereafter *a*-CDW and *b*-CDW, respectively.

Whether the CDW in cuprates has a quasi-1D stripe-like character or 2D checkerboard pattern has long been a controversy. The experimental fact that two CDW

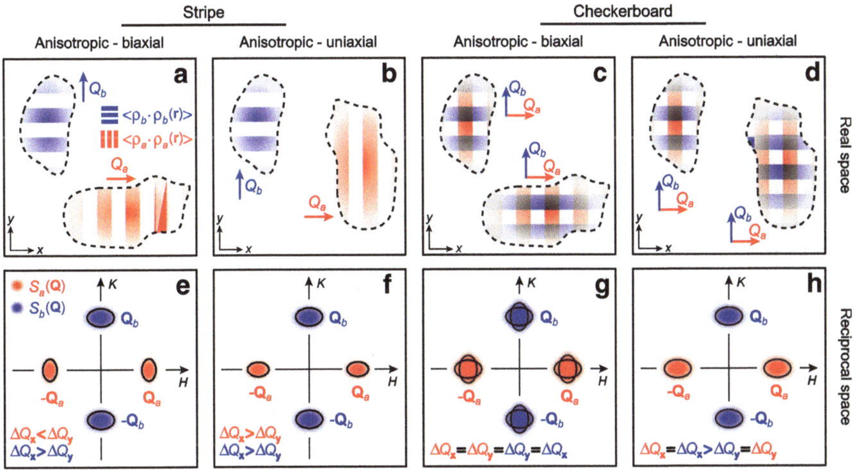

Fig. 5.6 Real space and reciprocal space pictures of different scenarios of CDW in cuprates. From [4]. Reprinted with permission from AAAS

peaks are observed at $\boldsymbol{Q} = (\mathrm{Q}^*\ 0)$ and $(0\ \mathrm{Q}^*)$ is not enough to exclude one of the two scenarios, since both scenarios are compatible with the presence of two CDW peaks along H and K. The experimental observation of a non-zero static structure factor at $\boldsymbol{Q} = (\mathrm{Q}^*\ \mathrm{Q}^*)$ can rule out the stripe scenario, but resonant x-ray elastic scattering experiments in $Bi_2Sr_{2-x}La_xCuO_{6+\delta}$ [3] have not detected any signal along $\boldsymbol{Q} = (\mathrm{Q}^*\ \mathrm{Q}^*)$ in the 2D-BZ. However, it still leaves a possibility that the null result is merely due to a low structure factor at $\boldsymbol{Q} = (\mathrm{Q}^*\ \mathrm{Q}^*)$, as the absence of the CDW intensity at higher harmonics of $\boldsymbol{Q}$, such as $\boldsymbol{Q} = (2\mathrm{Q}^*\ 0)$ or $(0\ 2\mathrm{Q}^*)$.

Comin *et al.* [4] summarized possible real space pictures of the charge density waves and the corresponding manifestations in the reciprocal space. These are shown in Fig. 5.6. The upper panels (Fig. 5.6a–d) of the figure show various real space scenarios of the CDW and their reciprocal space counterparts are shown in the lower panels (Fig. 5.6e–f). Semi-major/minor axes of an ellipse in the reciprocal space represent the HWHM of the CDW peak along H and K.

One important criterion to distinguish the checkerboard order from the stripe-like unidirectional order is that in the case of the former, the CDW peaks in reciprocal space should be equal in their peak widths/correlation lengths for the two orthogonal CDWs. That is, the a-CDW peak width along the a-axis should be equal to the b-CDW peak width along the a-axis and the same holds for the b-axis, as described in Fig. 5.6g, h. The measurements in underdoped YBCO with different doping levels do not satisfy this condition, thereby indicating we are dealing with the stripe scenario.

5.3.1 Confirming the Uniaxial Nature of CDW in Cuprates

The study of Comin *et al.* was successful in elucidating the anisotropy of the CDW correlation lengths, but the technical limitation in their experiment, which only allows rocking scans of the sample, made it difficult to accurately measure the transverse correlation length of the CDW. The ERIXS spectrometer that we used is equipped with a four-circle diffractometer in the vacuum chamber, which can overcome this technical difficulty.

Figure 5.7 shows elastic scans of both a- and b-CDW. For each orthogonal CDW, scans along H and K are compared. Our measurements clearly show that the scans along H and K for a single CDW are different in their peak widths, as the different lengths of the horizontal bars suggest. Both CDWs have larger peak widths along H than along K, indicating that they have a larger correlation length along the b-axis than along the a-axis.

Taking the widths along H and K of the CDW peak as semi-major/minor axes, the shapes of the CDW peaks in the 2D-BZ are drawn in the insets of Fig. 5.7. It is evident that the size of the CDW ellipses is different for a- and b-CDW. The size of the a-CDW ellipse is smaller than that of the b-CDW, which shows that a-CDW domains are on average larger than b-CDW domains. The unequal size of the ellipses of the a- and b-CDW further confirms the stripe scenario (Fig. 5.6b, f). We obtained similar results from 12 samples with the same doping level ($p = 0.12$). The minor variation between the measurements is reflected in the linewidth of the ellipses (Fig. 5.7 insets), which proves the high reproducibility of our results.

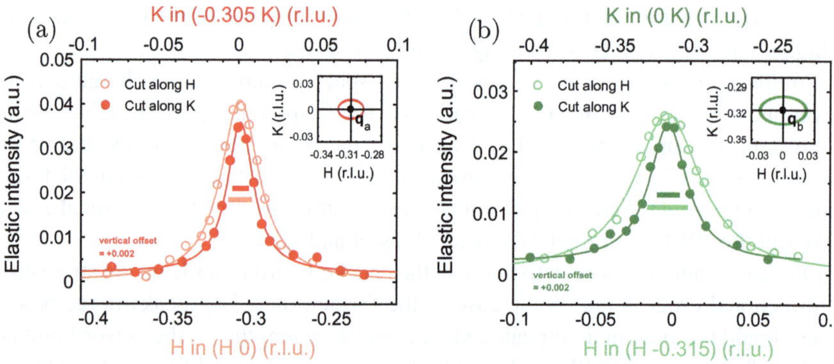

Fig. 5.7 **a** Scans of 2D-CDW modulating along the a-axis: cuts along H and K. Horizontal bars represent the HWHM (Half-width-half-maximum) of the peak with the corresponding color. **b** Scans of 2D-CDW modulating along the b-axis: cuts along H and K. Reprinted figure with permission from [1]. Copyright (2021) by the American Physical Society

5.3.2 Absence of the 3D-CDW Under b-axis Compression

Figure 5.8 shows how the three-dimensional correlation develops under uniaxial pressure along the *a*-axis. At zero strain, no clear peak is visible around integer L = 1. As the uniaxial pressure increases, elastic intensity around L = 1 starts to appear, but the intensity is at first rather weak, analogous to the IXS measurements. At the uniaxial strain level of −1.0%, a small, but clear elastic peak appears around L = 1. The temperature dependence of the new peak shows a maximum intensity at around T = 55 K, which is approximately the superconducting T_c at −1.0% strain along the *a*-axis, indicating its competition with superconductivity below T_c. The intensity of the L = 1 peak is not as strong as the peak observed at L = 7 in the previous IXS experiments, which might be attributed to the structure factor issues which is not completely known at present. These observations allowed us to attribute this as a signature of the uniaxial-pressure-induced 3D-CDW, successfully reproducing the previous result obtained with non-resonant inelastic x-ray scattering.

Our experiment provides moreover the opportunity to check for possible 3D-correlation along the *a*-axis under uniaxial pressure along the *b*-axis, which was not possible in the previous IXS experiments. In Fig. 5.9a, we show the strain dependence of the elastic scan along L around the CDW wave vector $\boldsymbol{Q}^a_{3D}$ = (0.31 0 L). Our measurement does not show any evidence of a 3D-CDW along the *a*-axis. It was previously observed that the intensity of the strain-induced 3D-CDW can vary dramatically at the superconducting transition. This necessitates additional measurements to check the presence of the 3D-CDW at several different temperatures.

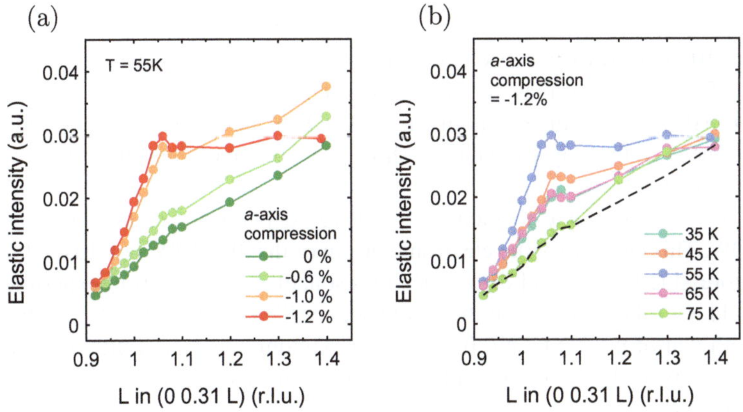

Fig. 5.8 **a** Elastic scans along L around the *b*-CDW wave vector $\boldsymbol{Q} = \boldsymbol{Q}_{\mathrm{CDW}}$ = (0 0.315 L) as a function of uniaxial pressure along the *a*-axis, at T = 55 K. **b** Elastic scans along L around the *b*-CDW wave vector $\boldsymbol{Q} = \boldsymbol{Q}_{\mathrm{CDW}}$ = (0 0.315 L) as a function of temperature, at −1.2% *a*-axis strain. The black dashed line represents the scan measured at unstrained condition, T = 55 K. Reprinted figure with permission from [1]. Copyright (2021) by the American Physical Society

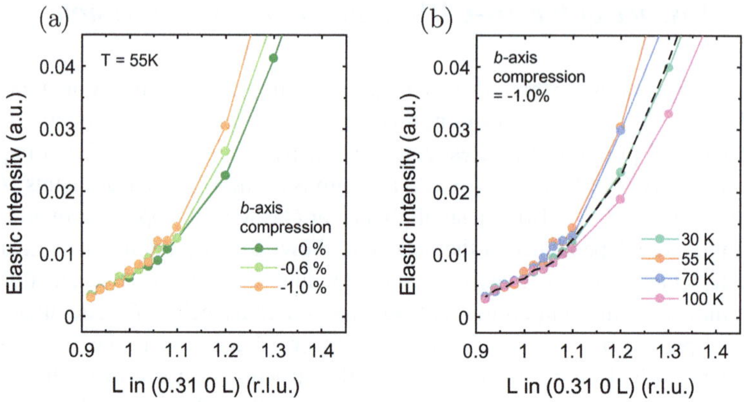

Fig. 5.9 **a** Elastic scans along L around the a-CDW wave vector $\boldsymbol{Q} = \boldsymbol{Q}_{\mathrm{CDW}}$ = (0.31 0 L) as a function of uniaxial pressure along the b-axis, at T = 55 K. **b** Elastic scans along L around the a-CDW wave vector $\boldsymbol{Q} = \boldsymbol{Q}_{\mathrm{CDW}}$ = (0.31 0 L) as a function of temperature, at −1.0% b-axis strain. The black dashed line represents the same scan measured at unstrained condition, T = 55 K. Reprinted figures with permission from [1]. Copyright (2021) by the American Physical Society

In Fig. 5.9b shows the temperature dependence of the scan along the L-axis. In the temperature range up to 100 K, no peak at around L = 1 was found.

5.3.3 *Same Resonance Profile of 2D- and 3D-CDW*

Figure 5.10a compares the resonance profiles of the 2D- and 3D-CDW. Previous studies [5] have shown that the 2D-CDW intensity has a resonance at the energy of the Cu(II) sites, i.e. the Cu atoms in the CuO_2 planes. This is reproduced in our measurement, as the empty red circles show a maximum of intensity at the pre-peak of the Cu-L_3 resonance.

We also measured the incident x-ray energy dependence of the strain-induced 3D-CDW. This can clarify the electronic origin of the 3D-CDW, but was not carried out before, due to technical difficulties in resonant x-ray scattering experiments at high field. Our data shows that the resonance profile of the 3D-CDW matches very well with that of the 2D-CDW. It shows a maximum intensity at the same incident photon energy as the 2D-CDW with the width of the resonance analogous to that of the 2D-CDW.

This indicates that both the 2D- and the 3D-CDW originate from the same $3d$ valence charges and the same electronic instability. This argument is further supported by the temperature dependence of the 2D- and 3D-CDW. Figure 5.10b shows how the 2D- and 3D-CDW intensities change as a function of temperature. Both of them have a peak in intensity at T$\sim T_c$ and lose their intensities below this temperature, exhibiting a competition with superconductivity that is naturally explained

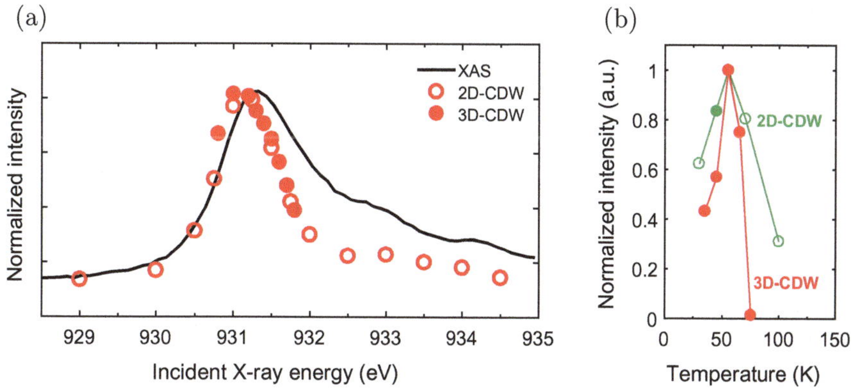

Fig. 5.10 **a** Elastic intensity of 2D- and 3D-CDW as a function of temperature. Filled(empty) green circles are the integrated elastic intensities of the $b(a)$-CDW peak. Red circles are the integrated elastic intensity in the range of L = [0.9 1.1] at different temperatures, −1.2% a-axis strain, subtracted by the unstrained scan at T = 55 K. **b** Comparison of XAS profile, incident energy dependence of the 2D- and 3D-CDW intensity. 2D-(3D-)CDW intensity was measured at the wavevector of $\boldsymbol{Q}_{2D}(\boldsymbol{Q}_{3D})$. Reprinted figure with permission from [1]. Copyright (2021) by the American Physical Society

by the same electronic origin of the two CDWs. Our measurement shows that the 3D-CDW emerges at T = 75 K at −1.2% a-axis strain and experiences a stronger competition with superconductivity than the 2D-CDW does, as illustrated as a sharper drop below superconducting transition at T∼55K.

The resonance profile of the strain-induced 3D-CDW is in sharp contrast to the observations made in epitaxial thin films [6]. The 3D-CDW observed in epitaxial thin films has a resonance that encompasses both Cu(I) and Cu(II) sites, which was interpreted as a contribution of the chain layer to the formation of the 3D-CDW. The absence of the oxygen chain order was claimed to be a factor that might enhance the nesting property of the Fermi surface to induce the 3D-CDW. In addition, it shows a negligible temperature dependence and exists already at room temperature. The distinct properties of the 3D-CDW in epitaxial films suggest a qualitatively different nature from the strain-induced 3D-CDW.

5.3.4 Anisotropic Response of the 2D-CDW Under Uniaxial Pressure

This section will show how a- and b-CDW change under uniaxial pressure along the a- and b-axis, respectively. For the CDW perpendicular to the uniaxial pressure direction, scans through $\boldsymbol{Q}_{\mathrm{CDW}}$ along both H and K were acquired. For the CDW parallel to the uniaxial pressure direction, only the scans along the modulation direction are presented.

5.3.4.1 *a*-axis Compression

Figure 5.11a shows an elastic scan of the *b*-CDW along the modulation direction K under *a*-axis compression. An increase of the *a*-axis pressure leads to a sharper elastic peak of the *b*-CDW with a larger intensity. This is consistent with the result of the previous chapter. The elastic scan along H, which is perpendicular to the modulation, also reveals a sharpening of the elastic peak, as shown in Fig. 5.11b. This indicates an increase of the transverse correlation length $\xi_{\perp}$ as well as the longitudinal correlation length $\xi_{//}$ of the *b*-CDW under *a*-axis compression. This result is summarized in Fig. 5.11c, where the elliptical shape of the *b*-CDW peak shrinks in the 2D-BZ. Between the unstrained condition and -1.2% strain, the longitudinal correlation length $\xi^b_{//}$ increases by almost 50%, whereas the transverse correlation length $\xi^b_{\perp}$ increases by 25%.

On the other hand, the response of the *a*-CDW under *a*-axis compression is quite different from the *b*-CDW response. Figure 5.11d shows an elastic scan of the *a*-

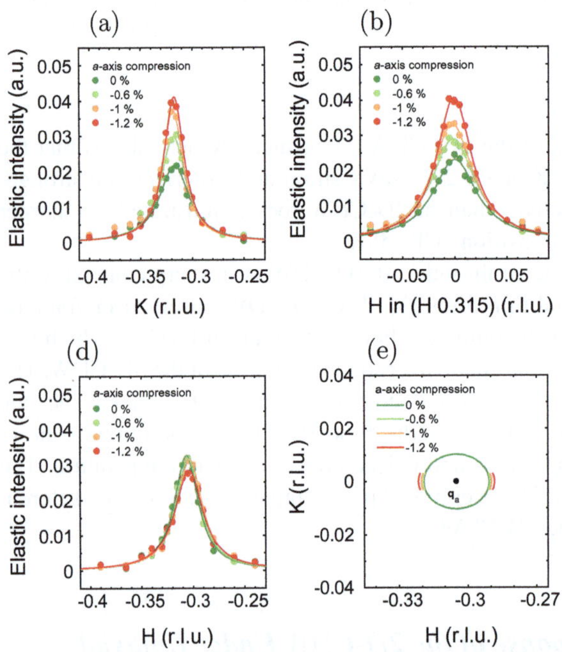

Fig. 5.11 Elastic scans of the *b*-CDW, cuts **a** along K and **b** along H under *a*-axis compression. In **a**, **b**, solid lines are the results of Lorentzian fits to the data. **c** Change of the peak widths of *b*-CDW under uniaxial pressure along the *a*-axis, visualized as an ellipse, where semi-major/minor axes are the HWHM of the elastic peak along H and K. **d** Elastic scans of *a*-CDW, cuts along H under *a*-axis compression. Solid lines are the results of Lorentzian fits to the data. **e** Change of the peak widths of *a*-CDW under uniaxial pressure along the *a*-axis, visualized as an ellipse, where semi-major/minor axes are the HWHM of the elastic peak along H and K. Reprinted figure with permission from [1]. Copyright (2021) by the American Physical Society

CDW along H under a-axis compression. Unlike the b-CDW, the elastic peak of the a-CDW does not show a noticeable change upon application of a-axis pressure. The width of the a-CDW peak along H is observed to show a slight increase upon a-axis compression, which is represented in Fig. 5.11e. The longitudinal correlation length $\xi^a_{//}$ shows a 15% decrease at -1.2% strain.

5.3.4.2 *b*-axis Compression

Figure 5.12a shows how the elastic peak of the a-CDW changes against b-axis compression. As the b-axis pressure increases, the a-CDW peak is enhanced and gets a bit sharper, indicating an increase of the longitudinal correlation length $\xi_{//}$. An increase of the transverse correlation length $\xi_{\perp}$ is also observed, as shown in Fig. 5.12b. This is summarized in Fig. 5.12c. The elliptical shape of the CDW peak gets smaller upon b-axis compression. Uniaxial strain of -1.0% along the b-axis leads to an increase

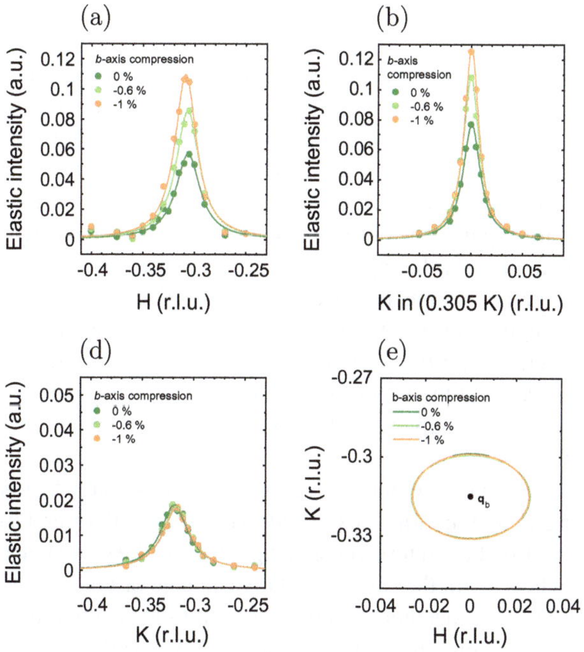

Fig. 5.12 Elastic scans of a-CDW, cuts **a** along H and **b** along K under b-axis compression. In **a**, **b**, solid lines are Lorentzian fit to the data. **c** Change of the peak widths of a-CDW under uniaxial pressure along the b-axis, visualized as an ellipse, where semi-major/minor axes are HWHM of the elastic peak along H and K. **d** Elastic scans of b-CDW, cuts along K under b-axis compression. Solid lines are Lorentzian fit to the data. **e** Change of the peak widths of b-CDW under uniaxial pressure along the b-axis, visualized as an ellipse, where semi-major/minor axes are HWHM of the elastic peak along H and K. Reprinted figure with permission from [1]. Copyright (2021) by the American Physical Society

of the longitudinal(transverse) correlation length $\xi^b_{//}(\xi^b_{\perp})$ by 12%(20%). The result is qualitatively symmetric and similar to the case of the b-CDW under a-axis compression.

As happened for the a-CDW under a-axis compression, uniaxial pressure along the b-axis barely affects the b-CDW. Figure 5.12d shows the elastic peak of the b-CDW as a function of b-axis pressure. It does not show any intensity change against pressure and also variations of the peak width remain negligible.

5.4 Discussion

5.4.1 Direct Versus Indirect Effects of Uniaxial Pressure

We first check the possibility that the observed 2D-CDW response against uniaxial pressure might not result from the direct coupling of the 2D-CDW to uniaxial pressure, but rather from an indirect coupling mediated by strain-induced changes of other properties.

Uniaxial pressure can, in principle, affect the effective hole doping in the CuO_2 plane. Charge transfer from the CuO_2 plane to the CuO chain layer is the key mechanism that determines the effective hole doping in YBCO. Hydrostatic pressure experiments [7] have shown that the decreased distance between the CuO_2 plane and the CuO chain layer is related to a valence change of Cu atoms in the CuO_2 plane. In-plane uniaxial pressure expands the out-of-plane lattice spacing and drives the charge reservoir of the chain layer farther apart from the CuO_2 plane (via quasi-elastic deformation), which generally results in a decrease of effective hole doping. In addition, a change in the strength of the superconducting order parameter can also influence the CDW. The competing, intertwined nature of these two orders requires that the enhancement of one order is at the expense of the other and vice versa.

Whichever scenarios of an indirect coupling of the 2D-CDW to uniaxial pressure we assume, they unanimously predict that both a- and b-CDW will change qualitatively in the same way under uniaxial pressure. For example, a-axis compression decreases the superconducting T_c and in turn, is expected to enhance both a- and b-CDW. The scenario of the increase of the interlayer spacing affecting the Cu valence change in the CuO_2 plane predicts again the same qualitative change of both a- and b-CDW (i.e. both increase or decrease together). These expectations are contradicted by our observation that, e.g., b-axis pressure enhances the a-CDW, while giving a negligible effect on the a-CDW. Therefore, we conclude that the observed effects of the 2D-CDW are a direct consequence of the uniaxial pressure on the 2D-CDW.

5.4.2 *b*-axis Compression Does Not Restore the C_4-Symmetry: Role of the CuO Chain Layer

At ambient pressure, $YBa_2Cu_3O_{6+x}$ has an orthorhombic crystal structure with $a = 3.83$ Å and $b = 3.87$ Å, due to the oxygen chains that run along the *b*-axis. In other words, in the absence of external strain, the CuO_2 plane is already under an orthorhombic distortion of $\sim 1\%$ with a broken C_4-symmetry.

High uniaxial strain of -1.0% along the long *b*-axis can bring the system to a nominal C_4-symmetric CuO_2 plane. Perfect C_4-symmetry is, however, not achieved due to the presence of the oxygen chain order. The data with high *b*-axis compression provide an effective way to study the role of the chain layer in the 2D-CDW phenomenology by minimizing the C_4-symmetry breaking effect. If the CuO chain layer plays a negligible role in the formation of the 2D-CDW, the C_4 symmetric CuO_2 plane should yield nearly degenerate *a*- and *b*-CDWs that appear as two types of identical domains with equal size rotated by 90°. However, no systematic evolution towards a formation of isotropic 2D-CDW domains under *b*-axis compression was observed. Rather, the large *a*-CDW gets slightly bigger in volume, while the volume of the *b*-CDW remains unchanged.

This is illustrated in the Fig. 5.13a. The ellipse sizes of the *a*- and *b*-CDW peaks are clearly different even in the limit of the strain-induced C_4-symmetric CuO_2 plane. In order to properly describe the 2D-CDW phenomenology, an inclusion of an additional parameter that couples with the CDW instability on the CuO_2 plane is therefore needed. Since the necessary condition for this outcome is some C_4-symmetry breaking, the only possible candidate remaining is the CuO chain layer. Several experimental studies have pointed out the role of the CuO chain layer in perturbing the quasiparticle behavior on the CuO_2 plane. Microwave conductivity

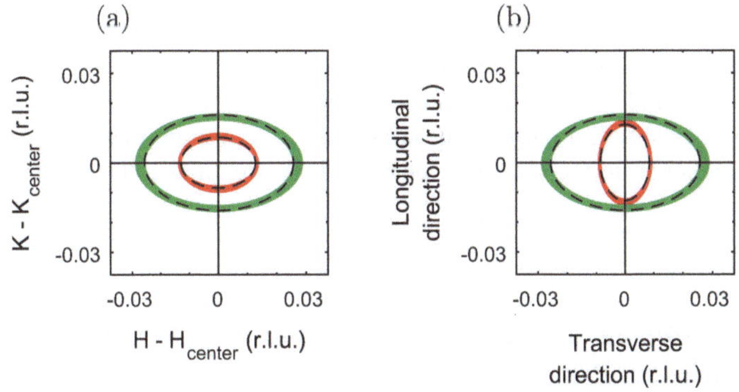

Fig. 5.13 The peak shapes of the *a*- and *b*-CDW at -1.0% *b*-axis strain **a** without rotation **b** with the *a*-CDW rotated by 90°. They were generated using the statistical average of the peak widths at the unstrained condition (Fig. 5.7 insets) multiplied by the observed correlation length change at the -1.0% *b*-axis strain. Red and green ellipses are from the *a*- and *b*-CDW respectively. Black dashed lines corresponds to the data measured at -1.0% *b*-axis strain, which are shown in Fig. 5.12

studies [8] have shown that disorder in the chain layer can act as a weak scattering source for quasiparticles in the CuO_2 plane. Oxygen disorder in the chain layer has also been discussed in an NMR study [9] as a pinning center for the formation of the 2D-CDW.

The influence of the CuO chain layer on the 2D-CDW can be investigated in more detail by comparing the longitudinal and transverse correlation lengths separately. This is justified by the quasi-one-dimensional stripe nature of the CDW. Figure 5.13b compares the peak shapes of both 2D-CDWs with the a-CDW rotated by 90°. The longitudinal peak widths of the two orthogonal 2D-CDWs show a small, but distinct difference, indicating the apparent role of the CuO chain layer. Quantitatively, the a-CDW has a larger longitudinal correlation length than the b-CDW by 15%. The observed difference could be attributed to a possible stronger disorder potential along the a-axis than the along b-axis, e.g. different types of disorder arising from broken segments of the oxygen ordering might pin the CDW differently.

Interestingly, a stark difference is observed in the transverse peak widths between the a- and b-CDW. The transverse peak width of the b-CDW is almost three times smaller than that of the a-CDW, which makes the observed difference in the longitudinal correlation length a minor effect. A possible explanation is that the transverse extent of the CDW domains might be mainly limited by the presence of defects in the chain layer. The correlation length of the oxygen order in the chain layer can give a good estimate on this effect. Although the direct measurement of these quantities are not available for our experiments, previous x-ray scattering studies [10] have shown that oxygen ordering reflections of YBCO with x = 0.62–0.77 typically have widths of $\Delta H \sim 0.03$–0.06 r.l.u. and $\Delta K \sim 0.007$–0.012 r.l.u. These values are roughly compatible with or at least proportional to the transverse peak widths of the CDWs, which are $\xi_{\perp}^{b} \sim 0.03$ r.l.u. and $\xi_{\perp}^{a} \sim 0.01$ r.l.u.

One simple way to test this hypothesis is to measure the transverse correlation lengths of the 2D-CDW before and after annealing to disturb the oxygen ordering in the chain layer. A measurement of the longitudinal correlation lengths using this method was carried out by Achkar *et al.* [11], showing no effect on the longitudinal correlation length. However, the role of oxygen order in determining the transverse correlation length has not yet been checked due to the few available resonant x-ray scattering setups with a sufficiently large angular coverage that enables a proper scan perpendicular to the scattering plane and thus remains to be verified in future experiments.

5.4.3 Anisotropic and Symmetric Response of the 2D-CDW to Uniaxial Pressure

Checkerboard charge order preserves the point group symmetry of the underlying crystal and in case of tetragonal lattice symmetry, the charge order should show C_4-symmetric responses to different directions of the uniaxial pressure, i.e. qualitatively the same response versus a- and b-axis compression, respectively. The comparison of the CDW parameters between the almost-tetragonal case of -1.0% b-axis strain and the unstrained case merely shows a small change. This provides a ground to approximate the system as almost C_4-symmetric. If we assume the checkerboard-type CDW in our system, qualitatively the same response versus uniaxial pressure is expected for uniaxial pressures along both in-plane directions. However, the observed responses of the 2D-CDW are highly anisotropic, which allows us to exclude the scenario of the checkerboard-type CDW.

The one-dimensional character of the stripe order that breaks the rotational C_4-symmetry naturally explains the observed anisotropic response of the 2D-CDW, since there is then no a priori reason that the uniaxial pressure response should be symmetric. The symmetric response of the 2D-CDW, with effects of uniaxial pressure on the a-CDW observed also on the b-CDW when rotated by 90°, can be seen as a reflection of the stripe-like unidirectional nature of the CDW.

5.4.4 Uniaxial Pressure Enhances the CDW Perpendicular to the Pressure Direction

Although the origin of the CDW enhancement under uniaxial pressure perpendicular to the modulation direction is not clear yet, our experiment indicates that it is an intrinsic property of the CDW instability in this material. Not only do both 2D-CDWs show such a feature, but also the 3D-CDW, which has the same wave vector as the 2D-CDW, is induced along the b-axis at the uniaxial strain of -1.0% along the a-axis.

Interestingly, it was theoretically proposed [12] that the CDW with a nematic form factor, namely a charge transfer between neighboring oxygens O(2) and O(3) in the CuO_2 plane, can arise from a multiorbital model of the cuprates with short-range antiferromagnetic correlations. NMR and x-ray scattering studies provide experimental evidence of inequivalent nearest-neighbor oxygens O(2) and O(3) in the CuO_2 plane. The nematic susceptibility naturally couples to the uniaxial pressure on the lattice and it might affect the CDW phenomenology by tuning the nematic order parameter, which subsequently affects the charge susceptibility. Further studies are necessary to elucidate the origin of the uniaxial pressure effect on the CDW.

5.4.5 Why the 3D-CDW is Observed Only Along the b-axis

The 3D-CDW was first observed by Gerber *et al.* [13] in $YBa_2C_3O_{6.67}$ under magnetic fields above 20 T. Further studies [14–16] revealed that it is observed in a similar doping range where 2D-CDW order is observed. In addition, they showed that the 3D-CDW is observed only along the b-axis and not along the a-axis and that the

Table 5.1 Comparison of the effect of the magnetic field and the uniaxial strain on the CDW in YBCO

Observation	Magnetic field	Uniaxial stress
Enhancement of the 2D-CDW	Yes	Yes
3D-CDW along the *a*-axis	No	No
3D-CDW along the *b*-axis	Yes	Yes
Incommensurability of the 2D- and 3D-CDW	Same	Same
Long-range ordering (Correlation volume > 10^4 unit cells)	Yes	Yes

enhancement of the 2D-CDW precedes the onset of the 3D-CDW. Many of the experimental features observed under magnetic field are analogous to the uniaxial pressure case, as described in the Table 5.1.

As regards the magnetic field effect on the formation of the 3D-CDW, Caplan *et al.* [17] proposed a classical effective field theory that qualitatively explains the observed CDW phenomenology, such as the half-integer CDW peak without field, the evolution to the peak at integer L, and subsequent changes of the CDW parameters under field. They argued that the main physics is driven by opposing forces, as also illustrated in Fig. 5.14.

(i) Coulomb repulsion $\tilde{U}$ between CDWs in the two CuO_2 layers *within* the YBCO unit cell that makes them out-of-phase. (different colors of CuO_2 planes in Fig. 5.14).
(ii) Defect potential V provided by the chain layer coupled to the CDW modulation that causes an in-phase coupling of CDWs in CuO_2 planes *across* different unit cells. (the same color of CuO_2 planes in Fig. 5.14).
(iii) A weaker Coulomb repulsion U between CDWs in CuO_2 layers *across* the chain layer.

The situation without the magnetic field is reproduced with the parameters $\tilde{U} \gg U \sim V^2$. The increase of the CDW amplitude and correlation length under high field leads to a dominance of the Coulomb repulsion over the defect potential and results in an in-phase inter-layer correlation while promoting an out-of-phase coupling between the CuO_2 layers across the chain layer, compared to the in-phase coupling in the absence of the field.

The experimental observation of the 3D-CDW only along the b-axis, not along the a-axis, suggests the presence of a stronger defect potential strength along the a-axis $V^a \gg V^b$ that pins the 2D-CDW and prohibits its 3D correlation. In this picture, if a sufficiently long correlation length of the 2D-CDW is achieved, the development of the 3D-CDW is not only possible along the a-axis in YBCO, but also in other cuprate compounds.

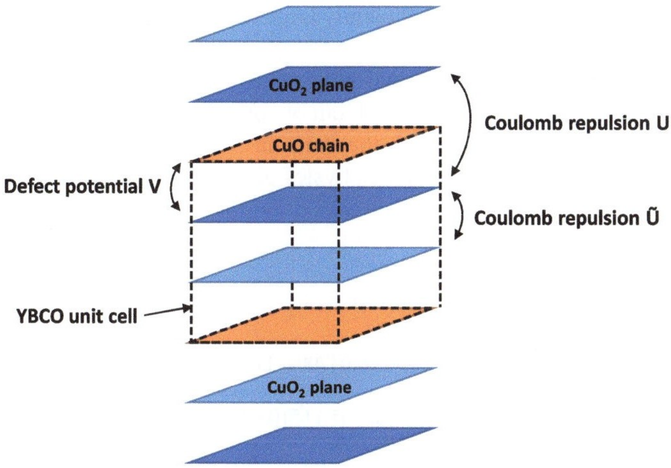

Fig. 5.14 Description of the parameters used in the classical effective field theory of the CDW proposed in [17]. The figure describes the situation without strain. Light and dark blue sheets represent CuO_2 planes. CuO_2 planes with the same color have CDWs in phase with each other. CuO_2 planes with different colors have CDWs out of phase with each other. Orange sheets represent CuO chain layers. Black dashed lines show the unit cell of YBCO

5.5 Conclusions

Detailed studies to elucidate the relation between uniaxial pressure and the CDW in $YBa_2Cu_3O_{6.67}$ were carried out using Cu-L_3 RIXS under uniaxial compression. Measurements of the correlation lengths of both a- and b-CDWs reveal a highly anisotropic response of the CDW to different directions of uniaxial pressure. This confirms the unidirectional nature of the CDW. In particular, a clear enhancement of the CDW under uniaxial pressure perpendicular to the CDW modulation is observed for both 2D- and 3D-CDWs. While a-axis compression induces the 3D-CDW, b-axis compression fails to induce a 3D-CDW along the a-axis, which is analogous to the high field effect on the CDW.

When the C_4 symmetry of the CuO_2 planes is forced back by b-axis compression, the CuO_2 plane provides a useful platform to shed light on the role of the CuO chain layer in the formation of the 2D-CDW. Our study shows that the CuO chain layer indeed affects both longitudinal and transverse correlation lengths of the 2D-CDW, particularly along the transverse directions.

In the framework of a classical effective field theory of the incommensurate CDW, our results can be understood as an interplay of the intra- and inter-layer Coulomb repulsion and the defect potential. The absence of the 3D-CDW along the a-axis implies that the defect potential provided by the chain layer is stronger along the a-axis than along the b-axis and prohibits the in-phase correlation along the c-axis. This picture indicates the possibility that 3D-CDW order can be realized in other cuprate systems provided that a sufficiently large correlation length of the 2D-CDW is achieved.

Our study demonstrates the potential of the uniaxial strain technique as an effective perturbation method in tackling the problem of correlated electron systems. Various experimental probes combined with the uniaxial strain technique are expected to provide new insights to understand the physics of intertwined orders.

References

1. Kim H-H et al (2021) Phys Rev Lett 126:037002. https://doi.org/10.1103/PhysRevLett.126.037002. Accessed 3 Jan 2021
2. Braicovich L et al (2014) Rev Sci Instrum 85:115104. https://doi.org/10.1063/1.4900959
3. Comin R et al (2015) Nat Mater 14:796–800
4. Comin R et al (2015) Science 347:1335–1339. https://doi.org/10.1126/science.1258399. https://science.sciencemag.org/content/347/6228/1335.full.pdf
5. Ghiringhelli G et al (2012) Science 337:821–825. https://doi.org/10.1126/science.1223532. https://science.sciencemag.org/content/337/6096/821.full.pdf
6. Bluschke M et al (2018) Nat Commun 9:2978
7. Jorgensen J et al (1990) Phys C: Supercond 171:93–102. https://doi.org/10.1016/0921-4534(90)90460-V
8. Bobowski J et al (2007) Phys C: Supercond 460–462. Proceedings of the 8th international conference on materials and mechanisms of superconductivity and high tem perature superconductors, pp 914–915. https://doi.org/10.1016/j.physc.2007.03.272
9. Wu T et al (2015) Nat Commun 6:6438
10. Zimmermann Mv et al (2003) Phys Rev B 68, PRB:104515
11. Achkar AJ et al (2014) Phys Rev Lett 113:107002. https://doi.org/10.1103/PhysRevLett.113.107002. Accessed 10 Sept 2014
12. Atkinson WA, Kampf AP, Bulut S (2015) New J Phys 17:013025
13. Gerber S et al (2015) Science 350:949–952. https://doi.org/10.1126/science.aac6257. https://science.sciencemag.org/content/350/6263/949.full.pdf
14. Jang H et al (2016) Proc Natl Acad Sci 113:14645–14650
15. Jang H et al (2018) Phys Rev B 97, PRB:224513
16. Chang J et al (2016) Nat Commun 7:11494. https://doi.org/10.1038/ncomms11494
17. Caplan Y, Orgad D (2017) Phys Rev Lett 119:107002. https://doi.org/10.1103/PhysRevLett.119.107002. Accessed 10 Sept 2017

Chapter 6
Conclusion

The non-resonant and resonant inelastic x-ray scattering experiments we have performed on $YBa_2Cu_3O_{6.67}$ ($p = 0.12$, $T_c = 65$ K) uncovered salient features of the CDW thanks to the extreme experimental conditions produced by uniaxial pressure, which were previously not accessible.

Uniaxial pressure along the a-axis not only enhances the existing quasi-2D CDW along the b-axis, but also induces the 3D-CDW along the b-axis. The 3D-CDW shows the same incommensurate period of modulation as the 2D-CDW, but is correlated in-phase between the neighboring layers, unlike the 2D-CDW that shows a very weak correlation with the next-nearest layer. The 3D-CDW has a much larger correlation volume ($\sim$52000 unit cells) than the 2D-CDW ($\sim$250 unit cells), revealing its long-range nature.

Phonon studies under uniaxial pressure show that the observed 3D-CDW transition coincides with the complete softening of an optical phonon. The observation of the soft phonon mode reconfirms the strong electron-phonon interaction present in high-T_c cuprates. Our result suggests that the momentum-dependent electron-phonon interaction might be responsible for the formation of the long-ranged 3D-CDW.

Comprehensive studies of the a- and b-CDWs under the a- and b-axis pressure revealed the anisotropic response of CDW to uniaxial pressure. Uniaxial pressure enhances the CDW perpendicular to the pressure direction, whereas it merely has a small effect on the CDW parallel to the pressure direction. This strongly anisotropic response of the CDW is not compatible with a checkerboard scenario of the CDW with C_4 symmetry and can only be explained with the unidirectional nature of the CDW.

The role of the oxygen chain layer in the formation of the CDW was demonstrated by removing the orthorhombic distortion of the CuO_2 planes using high b-axis compression and monitoring the CDW parameters. In the almost tetragonal CuO_2 planes under uniaxial pressure, the domain sizes of the orthogonal 2D-CDWs are observed

H. Kim, *Uniaxial Pressure Study of Charge Density Waves in a High-T_c Cuprate Superconductor*, Springer Theses,
https://doi.org/10.1007/978-3-030-99898-1_6

to be clearly different. This inevitably requires an important role of the oxygen chain layer in the CDW phenomenology.

A classical effective field theory of the CDW captures the key experimental features of CDW in YBCO with the interplay of the intra- and inter-layer Coulomb repulsion between CDWs in different CuO_2 layers and the defect potential in the chain layer that interacts with the CDW. In this picture, the 3D-CDW arises when the in-plane correlation length of the 2D-CDW gets sufficiently large and interlayer interactions overcome the defect potential. Our observation of strain-induced 3D-CDW only along the b-axis implies the presence of a stronger disorder potential along the a-axis. This mechanism of 3D-CDW formation opens the possibility of realizing 3D-CDWs in other cuprate systems, provided that a sufficiently long planar correlation length is achieved.

Uniaxial pressure provides a new platform to study the 3D-CDW observed in YBCO. The technical challenge of carrying out scattering experiments in the presence of magnetic fields exceeding 15–20 T makes it demanding to comprehensively study the 3D-CDW. The much easier implementation of the uniaxial strain device and the success of the IXS and RIXS experiments under uniaxial compression provide a strong motivation to carry out further studies using strain rather than field. In particular, combination of the uniaxial strain technique with various experimental probes, e.g. NMR, resistivity, Hall effect, optical spectroscopies, or heat capacity will give additional insights into the CDW nature, such as orbital symmetries, relation to the Fermi surface reconstruction, etc.

The investigation of the strain dependence of the CDW in other cuprate families will be also extremely interesting to address several questions, such as whether the observed anisotropic response of CDW is universal, and how the different nature of disorder present in various cuprate compounds affects the development of 3D-CDW long-range order. Comprehensive studies of different cuprate compounds are also expected to shed light on the theoretical description of the CDW phenomenology in cuprates, by e.g. by imposing constraints on the CDW symmetry.

Strain tuning can be utilized to elucidate the connection between the pseudogap and the CDW, e.g. a vestigial density wave order as a possible origin of pseudogap. Specifically, future experiments can monitor changes of the signatures of the pseudogap under uniaxial pressure, e.g. the pseudogap temperature T^*, the antinodal gap in ARPES, gap-like features in STM, optical spectroscopy, nematic order parameters from thermodynamic probes (ultrasound, torque magnetometry, Nernst effect), and compare the results with the response to strain of the CDW.

Strongly correlated electron systems are often characterized by a strong interplay of charge, spin and orbital degrees of freedom, resulting in nearly degenerate ground states. The thesis provides a new methodological approach that makes use of the uniaxial strain technique as a powerful tool to manipulate the complex energy landscapes of competing orders and uncover new physics in quantum materials.

MIX
Papier aus verantwortungsvollen Quellen
Paper from responsible sources
FSC® C105338

If you have any concerns about our products,
you can contact us on
ProductSafety@springernature.com

In case Publisher is established outside the EU,
the EU authorized representative is:
Springer Nature Customer Service Center GmbH
Europaplatz 3, 69115 Heidelberg, Germany

Printed by Libri Plureos GmbH
in Hamburg, Germany